T0093716

DARK
WINTER

RAINA MACINTYRE is a physician, epidemiologist and Professor of Global Biosecurity at UNSW and adjunct Professor at Arizona State University. She leads research in epidemic control, vaccinology and aerosol science. She is an expert in outbreak detection and mitigation, including that arising from bioterrorism and biological warfare. She has a 28-year track record in pandemics, epidemic infections, serious emerging infections, vaccines and control of respiratory viruses. She has worked as a clinician in hospitals, as an epidemiologist in a health department and as a researcher. In 2022, MacIntyre won the Eureka Prize for Leadership in Science and Innovation.

For my beloved Ayesha and Michael .

For those who lost their lives.

*For the courageous members of OzSAGE
and others who speak for the voiceless.*

*For Kelly-Ann, Nada and all the other health
workers who got infected at the front line and
were unsupported.*

*For David who spoke up for public health
and was punished for it.*

For all those who shine a light in the darkness.

DARK WINTER

An insider's guide
to pandemics and
biosecurity

Raina MacIntyre

NEWSOUTH

A NewSouth book

Published by
NewSouth Publishing
University of New South Wales Press Ltd
University of New South Wales
Sydney NSW 2052
AUSTRALIA
https://unsw.press/

© Chandini Raina MacIntyre 2022
First published 2022

10 9 8 7 6 5 4 3 2 1

A catalogue record for this book is available from the National Library of Australia

ISBN: 9781742237671 (paperback)
         9781742238487 (ebook)
         9781742239385 (ePDF)

Design Josephine Pajor-Markus
Cover design George Saad
Cover image Photo by Martin Sanchez on Unsplash

All reasonable efforts were taken to obtain permission to use copyright material reproduced in this book, but in some cases copyright could not be traced. The author welcomes information in this regard.

CONTENTS

PREFACE

CORONA DAWN

ON NEW YEAR'S DAY 2020, I WATCHED MY SON devouring his bacon and eggs at the Qantas lounge, while my daughter was immersed in her iPhone. I had been planning this holiday for the last 12 months, eager to splash out on my young adult children before they no longer wanted to holiday with me. This was to be their dream holiday, to make up for a lack of holidays for more than three years. Our last family vacation had been in July 2016 in Darwin. This time we were going to the United States. They were 18 and 20, and top of their wish-list was Orlando Disney World. Little did I know these were to be the last days of our old life, a life not consumed and changed forever by the pandemic. Nor would I have guessed that the US, ranked first in the Global Health Security Index for pandemic preparedness, would fail so badly or that Florida would be among the worst affected parts of the country, with casual workers driven to homelessness when Disney World closed its doors in 2020. Who could have guessed that in the span of two years the life expectancy in the US would drop by the same amount as a result of the COVID-19 body count? When the boarding call came, I scanned the news quickly on my laptop, aware that we would have a long flight without access to news. I spotted the news item about Wuhan: a small cluster of cases of an unknown pneumonia. The World Health Organization (WHO) was saying that it was not transmissible between people.

Just three weeks earlier I had been in the US running 'Pacific Eclipse', a biothreat simulation of an unknown epidemic that arises in the Pacific and becomes a pandemic. With a high-powered group of participants from the WHO, the US Centers for Disease Control and Prevention, US IndoPacific Command, UK bioterrorism experts from Porton Down and others, we exercised the war games of a pandemic in Washington, DC. The scenarios we created in the simulation exercise included stranded cruise ships full of infected passengers being denied safe passage in global waters, mass quarantine, travel bans, case finding and isolation of infected people, contact tracing, national interest overriding global health, and even a US election which impacted on the pandemic. Little were we to know that within months all of these elements would actually come to pass with COVID-19.

In the Pacific Eclipse simulation, the infection in question was smallpox. Although we had an effective vaccine, not all countries did. And those that had the vaccine held onto it, leaving the pandemic to spiral out of control in mega-cities in Asia. As a result, instead of being contained within a year, the pandemic went on and on for eight years. Smallpox, biological warfare and bioterrorism have been research interests for me since 2006. At the University of New South Wales (UNSW) I co-designed a course called Bioterrorism and Health Intelligence, to share knowledge gained from a career in pandemics and biosecurity with students. I have worked with police and military around biosecurity for many years now. This has given me a perspective of the vital importance of interdisciplinarity in biosecurity. The response to a pandemic – whether natural or unnatural in origin – requires so much more than health expertise. It usually requires health, defence, law enforcement, emergency services, occupational health, engineering and many other disciplines to work together. This was one of the objectives of Pacific Eclipse – to teach participants to walk in someone else's shoes and see the problem from a different perspective.

A couple of days after my family and I arrived in the US for our Disney World trip, the Wuhan story was still buried on page 3 and not yet headline news. I always keep an eye on outbreaks, aided by my epidemic observatory EPIWATCH, an artificial intelligence-driven system that scans news and social media for early signals of outbreaks. EPIWATCH has been my pet project since 2016. I've built it slowly, painstakingly, on a shoestring budget with deep knowledge of epidemics and an engine room of students. It is an artificial intelligence system but without funding or staffing, over the Australian summer holidays no one was present to check the signals it was spitting out. And, as we would later find out, the consequences of this were critical: had we been analysing the data at the time, we could have detected a signal for COVID way back in November 2019.

Almost as soon as my family and I got home, the COVID-19 epidemic exploded in China. It was clear there were many more cases than initially suspected, and that it was easily transmitted from person to person. Journalists immediately sought out my perspective on what was going on. For over 25 years I have been a regular voice of expertise in my field for the media, covering various outbreaks, among them the 2009 influenza pandemic. In early 2020 my phone was ringing hot from early in the morning until late into the night. I was getting calls from news agencies all over Asia, Europe and the US.

COVID-19 was upon us, ripping open many fault lines and exposing hidden truths about biosecurity. When the virus eventually hit, the impact in Australia was less than in many other countries. In a tactic reminiscent of the 1918 influenza pandemic, Australia used its unique island geography to keep the pandemic out for a whole year, shutting the international borders and buying time until vaccines were available – a honeymoon period of almost two years without facing the brunt of the pandemic as other nations did. In vaccination rollout and uptake, we lagged behind other countries. When the Delta wave hit in mid-2021, triggered by a failure to mitigate risk in airport transport, and compounded by a delay and lack of diversification in vaccine procurement, the Australian population was mostly unvaccinated. Still, we bought more time than other countries, enough to boast low death rates until the subsequent Omicron wave.

Many in Australia continue to cling to the glories of 2020, boasting about low mortality while simultaneously telling us the pandemic is over. By May 2022, the Australian Bureau of Statistics was already showing excess mortality from COVID, but our two-year period of grace – when Australia's international borders were shut – will not be fully reflected in excess deaths data until 2023. In the US, meanwhile, the impacts of COVID-19 have been massive. By 2022, life expectancy had dropped by a whopping two years. In the first six months of 2022 alone, the Omicron wave resulted in the deaths of over 8000 people in Australia – dramatically more than the 2200 or so deaths in all of 2020 and 2021. Hundreds of these were in younger adults and some were in children, and the deaths far exceed the national road toll each year.

In 2022 Australia saw supply chains affected, supermarket shelves empty, delays in essential services – all due to mass workplace absence. In the pre-pandemic period, somewhere between 2 and 5 per cent of workers may be off sick at any one time. At the peak of the Omicron wave, the figure was around 20 per cent. Mass cognitive dissonance is on display when people complain of chaos at airports all over the world, wondering why their luggage didn't arrive or their flight was cancelled. Part of the reason for such disruptions is that workers are sick with COVID-19. Vaccines on their own are not enough, and yet we have chosen not to use other layers of prevention, like ventilation, safe indoor air, masks, testing and tracing to mitigate the incidence of infection. Workplace absence, disruption to schools and households, hospitalisations and deaths are all a fraction of total case numbers. To reduce these, we must reduce transmission using a vaccine-plus strategy and ventilation.

Antivirals do exist, but there are not enough to be able to use them on a mass scale to add another layer of mitigation. Data are not yet available, but perhaps in the future rapid use of antivirals will cut the period of isolation and mean people can return to work sooner. They may even reduce the burden of long COVID. Yet to realise the promise of antivirals, testing is essential – they can never benefit the economy until testing is widespread, accessible and cheap or free. Giving people forewarning using digital tracing such as with QR codes will help. Meanwhile, good luck if you need to access health services or call an ambulance. Or if you get long COVID. In future, it is likely we will face a substantial burden of COVID-related chronic disease and disability. What this will do to our children is still unknown, but the available research suggests it is wildly reckless to sit by nonchalantly while the adults of tomorrow are infected en masse today, with the youngest still ineligible for vaccination.

The COVID-19 pandemic has brought with it disinformation, political meddling, counter-narratives and a flood of pseudo-experts willing to sell themselves for power and favour. It has also brought controversy about the origins of Severe acute respiratory syndrome coronavirus 2 (SARS CoV-2) – was it natural or unnatural? Did it arise from a lab leak at the high security lab in Wuhan? Determining the causes and patterns of epidemics requires detailed, careful analysis and expert knowledge. While even a child can tell if a building collapse is natural or unnatural – caused by an earthquake or a bomb blast – it's not so easy for epidemics. Even the most seasoned experts cannot easily determine the cause of epidemics. Why does knowing the origin of an epidemic or pandemic matter? Do community members want to know if their loved one died of a natural epidemic or as a result of human error or terror? Does it change accountability? Understanding the history of pandemics, lab leaks and biological warfare is critical to knowing why the origins of epidemics matter. Lab accidents are exceedingly common, and biological warfare has been used throughout recorded history. Why then would unnatural epidemics suddenly become rare today, when the technology is vastly more enabling and accessible than it was even ten years ago?

The challenges of ascertaining the origins of an epidemic is partly due to established ways of thinking about outbreaks, but denial of unnatural origins of epidemics is a recurring theme throughout history. Too often, the basic question of whether a disease arose from nature or from human error or terror is not even posed. When it is, it is usually agressively shouted down. This is partly due to protection of vested interests by scientists, and partly due to the desire of those in power to prevent fear in the masses or obfuscate their own involvement. Information warfare has become more insidious and powerful with the widespread use of social media. The COVID-19 pandemic has amplified all of these factors and exposed many truths. Using the lens of history, I show that not everything is as it seems.

1

BELIEVE THE UNBELIEVABLE

SEPTEMBER 1950. AS CALIFORNIANS WERE WAKING UP to sunlight playing over San Francisco Bay, unbeknownst to them, a hazardous sea spray was in the air. A biological aerosol spray was being pumped silently toward the California coast from a ship in the bay. Continuing for six days, the spraying originated from a navy minesweeper as part of a Cold War biological warfare experiment using bacteria. The US military thought the bacteria was harmless, but within days, a strange outbreak of urine infections occurred in San Francisco. Eleven people were admitted to Stanford Hospital with a serious and rare type of infection that turned the urine red. One of them, Edwin Nevin, died of a rare bacterial infection of his heart valves. The doctors who treated the patients wrote the outbreak up in a medical journal as interesting and rare, but never questioned whether it was unnatural. In medicine, we start from the assumption that every outbreak is natural. We are not taught in medical school or in public health training to question the origins of outbreaks, only to respond, treat and control them. But to Edward Nevin's family, it mattered whether their father died an unnatural and preventable death.

It would take 27 years and a lawsuit from Edward Nevin's son, for the US government to finally reveal the cause of this outbreak: Operation Sea Spray. Operation Sea Spray was a US Navy biological warfare experiment on US soil, affecting local communities. From a naval ship moored in the Pacific Ocean off the San Francisco coast, the US Navy had released enough aerosolised bacteria *Serratia marcescens* and *Bacillus globigii* to infect nearly all of the city's inhabitants. Taking samples, the military found that the bacteria had spread all over the city and suburbs beyond, from San Francisco to Sausalito, Albany, Berkeley, Oakland, San Leandro, Daly City and Colma. The residents of these cities would have inhaled millions of bacterial spores. Health authorities, however, were never alerted to the dangerous experiment. It is thought that this experiment may have also resulted in cases of pneumonia and heart valve infections, as well as permanent, long-term changes to the microbial ecology of the region.

Operation Sea Spray was conducted during the Cold War. It was one of several experiments into biological warfare conducted by governments on their own populations. The Cold War was a time when an arms race in biological weapons was underway, not only in the US and the Soviet Union, but also in the UK and other countries. Each country thought nothing of experimenting on their own people in the pursuit of military supremacy. The government seeing the masses as inconsequential, or even expendable is nothing new.

In the UK, similar experiments, including the 'Sabotage Trials' and DICE Trials, were conducted over communities between 1952 and 1975. In these cases, aircraft were used to spray bacteria over large areas without the knowledge or consent of the residents. Part of the larger Dorset Biological Warfare Experiments, these tests started with the spraying of the fluorescent chemical zinc cadmium sulfide, which accumulates in biological systems and can cause kidney, bone and respiratory problems with long-term exposure. Many families would later also report birth defects in children born in Dorset around the time of these experiments. Cadmium may also cause lung cancer. The US conducted similar experiments with zinc cadmium sulfide, and found that spraying over the mid-west of the country had widespread effects, with particles detected as far away as New York.

By the 1960s, the UK military expanded the DICE Trials to include the testing of bacteria such as *Bacillus globigii* and *Escherichia coli* (E. coli), with the aim of seeing how effective aerial spraying might be for biological warfare. They found that a single aircraft could infect an area of 16 000 square kilometres. Other experiments carried out on the unwitting populace in the UK and the US included releasing *Bacillus globigii* in subway train networks to see how far it would spread with the air currents generated by the movement of trains. As a bacterium, *Bacillus globigii* is similar to anthrax but thought to be safe. In 1966, in the middle of rush hour in the New York subway, the US Army dropped lightbulbs filled with more than 87 trillion *Bacillus globigii* bacteria in order to understand the behaviour of anthrax aerosols in the train tunnels. They wanted to determine how far the bacteria would travel and how easily it could be decontaminated. At the time, *Bacillus globigii* was thought to be harmless to people.

Today it is known to cause infections in the blood stream and lungs, as well as meningitis in people with underlying medical conditions. During the New York experiment over a million commuters were exposed, but no data were collected on illness after the event. And no New Yorker who might have fallen ill was aware that they had been guinea pigs in a government-led bioweapons experiment.

Experiments conducted by governments on their own populations can be traced back even further. During the Second World War, the US tested chemical weapons on African American, Puerto Rican and Japanese American soldiers, using white soldiers as controls. The aim was to determine if there were racial differences in chemical injuries. The soldiers were put in chambers and exposed to Mustard Gas and other chemicals. Importantly, however, none of these experiments were documented on the soldiers' official service records. As a result, this meant that some 60 000 soldiers subjected to deliberate chemical exposure were not eligible to receive medical care or compensation for the lifelong chronic injuries they suffered as a consequence. Some, however, did receive commendation certificates for 'subjecting themselves to pain, discomfort, and possible permanent injury for the advancement of research in protection of our armed forces'. African American, Puerto Rican and Japanese American troops were considered 'lesser' and untrustworthy, were segregated and generally not allowed to have weapons, but assigned menial tasks. In 1993 the experiments were declassified, but it was impossible to track all the victims, because official records were never kept. Eventually, in the 1990s, surviving victims were paid benefits. But only 193 out of about 2000 claims received compensation. More recently these events are depicted in the brilliant Amazon Prime series *Them* which uses the horror genre to show the lives of an African American family who move into a white neighbourhood in Los Angeles in the 1950s. It touches on the chemical weapons testing on African American soldiers.

During the Second World War the UK similarly experimented with biological and chemical warfare. In 1943 a biological weapons plan, with the aim of releasing anthrax in Germany, began. The British government purchased Gruinard Island off the Scottish coast as the test site. Sheep and cattle were placed on the island as the test subjects, and teenagers were hired to help with the experiments. At least 22 anthrax bombs were made to rain on Gruinard Island and all the animals were observed to die within three days. Worse still, the carcasses of infected animals washed ashore, infecting livestock on the mainland. The government further exacerbated the effects by collecting dead animals, putting them in caves and then using explosives to get rid of the evidence – this backfired and resulted in anthrax being disseminated onto the mainland from the blasts. Gruinard Island was assessed to be so highly contaminated that it would be uninhabitable for at least 100 years. Documents detailing what really happened there were not declassified for more than 50 years. By 1987, however, after decontamination, the island was deemed fit for habitation. After the Second World War, British scientists went even further by conducting open sea testing of anthrax in the Caribbean. It was not until years later that anyone had any clue about these and other experiments, following their disclosure by governments. This demonstrates the tendency of governments to hide shocking secrets, as well as the inability of health experts to recognise unnatural outbreaks occurring in plain sight.

A very unusual outbreak

Being able to identify and analyse unnatural outbreaks is essential for the effective management of epidemics. This critically important point is one that I always drive home to my students. In a course I designed at the UNSW entitled 'Bioterrorism and Health Intelligence', I take students through an outbreak of salmonella in the US. The exercise uses as a case study a real outbreak, and I provide sequential data on it, along with the conclusions of the Centers for Disease Control and Prevention and local public health authorities. I ask the students to analyse the outbreak and make their own conclusions. The circumstances are as follows: in 1984 hundreds of people in a city in Oregon started getting sick with gastroenteritis. All the patients had eaten in at least ten different restaurants. I walk the students through the investigation. We look at the pattern of foods eaten by those who did and did not get sick, but there is no common ingredient or food – not the eggs, or the milk, the water supply or any other product. The inspection of affected restaurants show sanitation, water and food preparation to have been reasonable. All of the restaurants had a salad bar, and in ones that served the same food in private dining rooms, only people who ate from the public salad bar got sick.

The outbreak strain of salmonella was an unusual one, quite different from other salmonella outbreaks seen in

Oregon. With over 700 cases, it was the biggest outbreak the US had seen to date. Local health authorities called in the Centers for Disease Control and Prevention for assistance. The outbreak experts concluded that unsanitary food handling was the cause. This is not surprising. After all, anyone who works in outbreak investigations knows that salad bars are notoriously unsanitary and often associated with outbreaks. As a trainee field epidemiologist with the Victorian Department of Health and Human Services in the 1990s, I investigated many salad bar outbreaks myself. The inside knowledge I gained during those investigations make me wary of ever eating from a salad bar. The circumstances were often stomach turning. In one case, potato salad was revealed to have been contaminated with faecal bacteria. In another famous outbreak in the US, a disgruntled kitchen hand who had Hepatitis A but was asymptomatic, urinated into the coleslaw as revenge, causing an outbreak of the virus.

In Oregon, following the salmonella outbreak, the restaurants were shut and the outbreak ended. Case closed. At this point in the case study exercise, the students mostly tend to agree with the Centers for Disease Control and Prevention about the origin of the epidemic. I ask them why they agree, when the inspections had shown that sanitation in these establishments was okay. Were the cooks unsanitary in all ten restaurants? What else could explain an outbreak affecting ten different restaurants, I ask? Why did people eating the same food in private dining rooms not get sick? Our discussions touch on the particular response of politician Jim Weaver, who insisted at the time that it was a bioterrorism attack by a local cult. He was laughed at and shouted down by the Centers of

Disease Control and Prevention and the public health experts in Oregon. Convinced, Weaver went to the media with his theory, and was ridiculed by the public health folk and labelled a conspiracy theorist. Most of the students in my class agree that Weaver was a crazy conspiracy theorist. Anything outside our accepted reality or knowledge base must be a conspiracy, surely? And yet the facts before us are clear: ten different restaurants affected and sanitation and water were fine. There was no common ingredient that could explain the outbreak. Deliberate contamination is clearly a possible explanation, but why is it so hard to see it? And why is anyone like Weaver, who suggests the possibility ridiculed and shouted down?

Whenever I teach my Bioterrorism and Health Intelligence course, I always remind students that in order to detect unnatural outbreaks, you must first ask the question 'Is it natural or unnatural?' The reason that we need to know if epidemics are natural or unnatural is that it determines the course of action. In both cases, public health measures are put into action, but in the case of an unnatural outbreak, further specific action may be needed, including finding the perpetrator (or perpetrators) in order to prevent further attacks and bring them to justice. About six months after the Oregon salmonella outbreak, cult leader Bahgwan Sri Rajneesh admitted that his cult had poisoned the restaurants. But no one believed him. This reluctance surely illustrates a hard-wired recalcitrance to accepting the possibility of unnatural outbreaks. It begs the question: how hard is it for health experts to recognise an unnatural outbreak when even a confession is not believed? In the end, evidence confirmed Rajneesh's confession. Quite by accident, a year after the outbreak, while conducting an immigration raid, the FBI stumbled on the Rajneesh lab on their sprawling ranch. The cultures of salmonella found in the lab were determined to be an exact match to the rare outbreak strain. That was the final evidence – forensic evidence collected by the FBI, without a single contribution from health experts.

This failure in public health was not due to a lack of availability of technical experts. Indeed the US Centers for Disease Control and Prevention arguably boasts the greatest concentration of technical experts in the world. But technical expertise without the ability to break the shackles of dogma, ideology and group-think is as good as useless. It was not until a full 12 years after the event that the findings from the Rajneesh case were talked about and made public. My students, just like the public health officials at the time, cannot think of a reason why anyone would choose to engage in such an act of bioterrorism, and so cannot believe it. Occasionally, a student analysing the outbreak on the face value of the facts will come to the conclusion that 'it's unnatural', but the majority of students do not. They cannot think of a motive and cannot imagine the unimaginable. In the end, the actual motive turned out to be political. The Rajneeshis were in conflict with the local government about a land issue and were running their own candidates in the local election. They had tried to register their 4000 or so cult members to vote, including homeless people they had collected off the streets of Portland, but this backfired and was disallowed. So their plan B was to poison the town water supply to make the locals too sick to vote on election day, thereby winning the election and taking control of the town. The restaurant attack was their practice run.

The Rajneesh salmonella attack in Oregon in 1984 exemplifies the failure of public health when it comes to recognising unnatural outbreaks. All of the elements – the failure to even ask the question as to whether the cause was natural or unnatural, the overriding of clear red-flag evidence with beliefs, and the labelling of anyone who suggested it was an unnatural cause as a conspiracy theorist – prove how hard it is to expose the origins of an unnatural outbreak. The Rajneesh case showed also that even a confession (which is a rare occurrence in both deliberate attacks and lab leaks) is not enough for health authorities. Given these barriers, what hope then is there for ever detecting unnatural outbreaks?

The problem is that the investigation of unnatural outbreaks requires the skills of police or intelligence analysts, not health experts. There is little collaboration between disciplines, except in some notable cases such as the US anthrax attacks in 2001. In that case, when it became clear the strain of anthrax had likely come from a US military laboratory, the FBI and Centers for Disease Control and Prevention collaborated in a joint investigation. In contrast to the Rajneesh salmonella attack, however, the outbreak was clearly unnatural, caused by anthrax being mailed in envelopes complete with threatening letters purporting to be from terrorists to several people shortly after the September 11 Twin Towers attack. In terms of ascertaining origins, however, the most challenging outbreaks are the ones that spread from person to person. This is because once an outbreak starts spreading, a highly contagious disease pattern will look the same whether the origin was natural or unnatural. The Rajneesh case was not spread from person to person. Rather it was what we term a 'point source outbreak'.

The origin of such cases is usually a common source such as food, water or the environment. In the Rajneesh case, it should have been easy to pick the outbreak as unnatural, but ingrained biases and lack of training produced a barrier to effective and accurate identification of causation. In order for transformational change to occur, we need public health experts instilled with forensic and criminal investigation skills as undergraduate students, so that if they ever encounter the health effects of an unnatural outbreak, they are able to think outside the square.

It's raining anthrax

Historically, very few unnatural outbreaks have been correctly recognised as such at the time. A classic example of this is the notorious lab leak from the Soviet bioweapons facility at Sverdlovsk (now Yekaterinburg), east of the Ural mountains. One night in 1979, so the story goes, a technician at the Soviet bioweapons facility forgot to replace a valve in the ventilation system, resulting in weaponised anthrax being accidentally pumped out into the town all through the night. Other reports suggested there had been an explosion at the bioweapons factory and thousands had died. The official death toll was closer to 100, but to this day no one knows the true number of deaths. People died of inhalational anthrax up to 4 kilometres away from the site and by some reports as far as 50 kilometres away, showing how far anthrax spores could disperse in the wind.

Anthrax is a bacterium which has three main forms: inhalational, cutaneous or gastrointestinal. Naturally occurring anthrax (caught when people handle contaminated meat or animals) is usually cutaneous, presenting with a black lesion on the skin. Untreated, it can progress to bloodstream infection and be fatal – between 10 and 40 per cent of cutaneous infections may result in death, but this drops to less than 1 per cent with antibiotic treatment. Anthrax is endemic in many countries, in livestock, wild animals and also in the soil. Over 95 per cent of naturally occurring anthrax is cutaneous. On encountering a case of inhalational anthrax, an unnatural cause should be suspected, as this form is exceedingly rare in nature. Anthrax forms spores, which are an inactive form of the bacteria that can lie dormant for many years but can activate when they come into contact with the human body and then begin reproducing and forming bacteria, which causes the illness. Inhalational anthrax arises when people inhale airborne anthrax spores which causes a rapidly fatal illness. Unlike SARS-CoV-2, anthrax is not contagious from person to person.

As a result, the Sverdlovsk incident remained localised. Kept under wraps by the Soviet government until 1992, the Sverdlovsk incident was suspected as unnatural by US intelligence agencies when satellite imagery and signals intercepts suggested the area was under military control at the time, with road closures and other signs of a major incident. It looked like a serious national security threat had occurred, but the Soviet authorities denied everything, asserting that the outbreak was natural. Interestingly, in early October 2019 there was evidence of a similar kind in Wuhan. Reduced road traffic, roadblocks around the Wuhan Institute of Virology (WIV) and loss of phone signals around WIV suggested a major incident had occurred, estimated to have taken place between 6 and 11 October 2019, according to a leaked US intelligence report titled 'MACE E-PAI COVID-19 Analysis'.

In 1979, whilst the US intelligence agencies were convinced based on satellite and other intelligence that there had been a lab accident in Sverdlovsk, their leading biosecurity and bioweapons expert, Professor Matthew Meselson, a renowned geneticist, visited Moscow to investigate. Meselson agreed with the Soviets that it was a natural outbreak of anthrax. This may have seemed reasonable to him based on some of the available evidence – after all, there was naturally occurring anthrax in the region, there were infected cattle during the outbreak, and the Soviet authorities claimed that cattle had been infected before the human outbreak. Other scientists and medical experts also disagreed with the US intelligence agencies, displaying the typical reluctance to accept unnatural origins for outbreaks of disease. Leading science and medical journals accordingly published 'natural origins' papers on the incident. The Soviets, meanwhile, made outraged denials of a lab accident. They rejected allegations of an offensive biological weapons program, and even accused the US of amplifying the arms race and Cold War tensions.

The disconnect between the US intelligence agencies and scientists reflects their different skills sets and a different standard of proof around the question of origins. It also reflects the natural vested interest that scientists have in rejecting laboratories as a source of epidemics or even pandemics. After all, if this were the case, it could result in increased regulation and decreased freedoms for scientists. This pattern of opposition between law enforcement and science has been repeated in other cases, from Bruce Ivins – the US Army

Scientist and main suspect in the 9/11 anthrax attacks – to US plague scientist Thomas Butler, sentenced to prison time in 2004 for multiple charges relating to his research and other matters. In both cases, which will be discussed later, groups of scientists wrote outraged letters defending their colleagues and pillorying law enforcement officials for investigating and prosecuting them. When I have talked to law enforcement officials about those cases, however, they have shrugged and said 'I don't care who he is. He did the perp walk just like any other person who broke the law.'

In the case of Sverdlovsk, the Soviets put on a terrific theatrical performance around 'proving' it was a natural outbreak. They culled animals and killed stray dogs at the markets, pointing to these sites as the source of contaminated meat and to the animals as responsible for spreading it. When I heard of the wet market being the focus of the Wuhan outbreak in 2019, I immediately thought of Sverdlovsk. For most of 2020, medical experts kept pushing the wet market theory, even though half of the first recognised cluster of cases, including the first two cases, had not visited the market. After the lab leak theory gained fresh traction in 2021, a counter-narrative of new wet market theories started doing the rounds and were given an airing in top scientific journals. In Wuhan, Chinese authorities also closed down the market, culled animals and did extensive testing around the immediate area. No bats had been sold at the market, which was mainly a seafood market, and traces of the virus were not found in the wild animal sections, but rather in the seafood sections. This then generated a narrative about frozen foods being the culprit. Meanwhile, as early as February 2020, scientists began publishing adamant claims that SARS-CoV-2 had a natural origin from animals. Anyone who dared question the origin was shouted down as a conspiracy theorist.

In early 2020 we barely knew anything about the virus. How then could any scientist be so certain about its origins? Having researched the origin of outbreaks for many years before the COVID-19 pandemic and having studied past unnatural outbreaks, I saw many red flags about the origins of the COVID-19 virus. But, seeing how viciously others were being attacked, and having been attacked in 2014 for suggesting MERS-CoV could be explained either by recurrent animal introductions or by an unnatural epidemic, I stayed quiet. The first WHO investigation into the origins of COVID-19 continued the wet market narrative, with an implausible hypothesis about it spreading through frozen food. This was long after it had spread globally from person to person and caused the most catastrophic pandemic of our lifetimes. Food was clearly not the main way it spread. In both the Sverdlovsk and SARS-CoV-2 outbreaks, initial reports attributed them to animals in the markets, with scientists zealously spruiking natural origins. A flood of narratives and counter-narratives then followed in what is essentially an information war.

In 1992, after the fall of the Soviet Union and 13 years after the Sverdlovsk accident, President Boris Yeltsin came clean about the incident. Meselson and the flood of papers published in scientific journals had been wrong. An accident had indeed taken place in the Soviet anthrax bioweapons facility. The cows infected prior to the outbreak of human infections had resulted from testing that involved scattering anthrax on the ground and observing the impact on animals.

The Soviets had made use of a whole island (named Aralsk-7) in the Aral Sea for testing bioweapons on animals. Soon after the revelations, Meselson returned to Russia to do a follow-up investigation. In 1994, the findings were published in the prestigious journal *Science* in an article co-authored with his wife Jeanne Guillemin. The article outlined the impact of a deadly plume of anthrax emanating from Compound 19 in the bioweapons facility to the adjacent town. Meselson and Guillemin still gave the Soviets the benefit of the doubt, suggesting it was less than a gram of anthrax, consistent with defensive, rather than offensive biological warfare. At the time, US military anthrax experts disagreed with this assessment, suggesting the accident involved much larger quantities of weaponised anthrax (perhaps as much as 10 kilograms), with the implication that the Soviet Union was involved in offensive bioweapons research in breach of the 1972 Biological Weapons Convention which prohibited such research.

The suspicions of US military anthrax experts were supported by Colonel Kanatzhan Alibekov, who confirmed that the Soviet bioweapons facility had made 300 tons of anthrax in less than a year. Alibekov, a Soviet defector who would later be known as Ken Alibek, was deputy director of the Soviet bioweapons program and had a deadly weaponised strain of anthrax named after him (*Alibekov anthrax*). He exposed the Soviet biowarfare program in his book *Biohazard*, revealing how the Soviets were brazenly breaching the Biological Weapons Convention and operating a network of labs across the country, including the smallpox lab Vector, Aralsk-7 and Stepnogorsk, which took over as the main anthrax facility after the Sverdlovsk accident.

Meselson is often credited with uncovering the truth about the Sverdlovsk outbreak. Meselson, however, after initially agreeing with the Soviets that it was a natural outbreak, minimised the incident to a small leak. With the passage of time, this history has been forgotten. In reality it took Meselson over a decade to agree with intelligence agencies that this was an unnatural outbreak. And even that was prompted only by a confession by Yeltsin.

So what we see here is a recurrent pattern of failure to recognise the facts, a denial in the face of facts, or a minimisation of serious unnatural outbreaks. From Operation Sea Spray and the Rajneesh attack to Sverdlovsk, all of the responses to these incidents have had common characteristics. We also see a recurrent pattern of truth being suppressed for a very long time after the incident. In all three cases discussed in this chapter, it took many years for the truth to come to light – up to 27 years in the case of Sea Spray.

2

INSIDER THREAT

IN SEPTEMBER 2001, IN THE WAKE OF 9/11, THE UNITED States faced another attack. Weaponised anthrax was mailed out to multiple targets, including senators Tom Daschle and Patrick Leahy, and several media outlets. The first case and death was in Florida: Robert Stevens, a photo editor of *The Sun*. This was followed by more deaths in New York and Washington DC. Seven anthrax-laced letters were thought to have been mailed from parts of New Jersey and Florida where Arabic-speaking communities lived. Some of the 9/11 hijackers were also from these areas. The New Jersey letters were postmarked 18 September 2001, a week after the World Trade Center attacks and sent to the *New York Post* and *NBC News*. The letters had the message: '9-11-01, THIS IS NEXT, TAKE PENACILIN [*sic*] NOW, DEATH TO AMERICA, DEATH TO ISRAEL, ALLAH IS GREAT.' A series of hoax and copycat letters also followed, throwing the United States into chaos. A subsequent letter to Senator Tom Daschle read '9-11-01, YOU CANNOT STOP US. WE HAVE THIS ANTHRAX. YOU DIE NOW. ARE YOU AFRAID? DEATH TO AMERICA. DEATH TO ISRAEL. ALLAH IS GREAT.'

In the end, the 2001 mail attack caused five deaths and 17 non-fatal cases of anthrax, shut down the US Postal Service and the US Congress, and required extensive decontamination efforts in multiple locations including government buildings. Initially, influential op-eds by former CIA director James Woolsey and Australian diplomat Richard Butler pointed the finger at Iraq, despite no evidence supporting this theory. This provided fuel for the building momentum for the US to wage war on Iraq. However, genetic sequencing of the anthrax came up with an unexpected finding: it was not a strain from Iraq, or even from overseas. It was, in fact, a domestic US strain, thought to be the deadliest natural anthrax strain in the world, from a cow that had died in Texas in 1981.

The strain of anthrax used in the attacks was the Ames strain, a US strain that had been used in defensive biological weapons research at US Army Medical Research Institute of Infectious Diseases (USAMRIID) at Fort Detrick, Maryland. Radiocarbon dating put the spores to be no more than two years old, and forensic examination showed some of the letters contained modern, weaponised anthrax. Some speculated that the attack could have originated from other US labs which used the Ames strain. The US also shared the Ames strain with the UK biodefence facilities at Porton Down, so the strain was present outside the US.

More highly weaponised and concentrated anthrax was found in the letter sent to Senator Daschle compared to the *New York Post* letter. Weaponisation involves spray drying and coating with substances such as silicone to make the spores more airborne and easier to disperse widely. The *New York Post* letter was reported to contain a brown substance with less concentrated spores, whilst the Leahy letter contained white powder and the most advanced form of weaponised anthrax spores. At the time, the coating on the spores in the Leahy letter was a mystery to FBI investigators, and reportedly could not be reverse engineered to identify what it was. USAMRIID microbiologist Peter Jahrling initially assessed the Leahy strain as highly weaponised, but some years later in 2008 backtracked and claimed he had made a mistake. FBI and outside scientists also reported they found silicone, but concluded the anthrax was natural. Since it is unlikely that natural anthrax contains high amounts of silicone, there are inconsistencies in the story. There was also tin and other bacterial elements in some of the samples. In 2008 Senator Leahy maintained that he believed the attack was too sophisticated to be perpetrated by a lone wolf, and that a group of people would have to have been involved. A former attorney, Leahy was given more access to material about the case than others, and said it was 'an old prosecutor's instinct'. Today, Leahy remains the senior senator from Vermont and is the most senior member of congress.

At the time of the FBI's investigations, the main suspect was a Fort Detrick scientist, Dr Bruce Ivins. A microbiologist and anthrax expert with patents on anthrax vaccines, Ivins was initially advising the FBI in their investigations. The US Commission on the Prevention of Weapons of Mass Destruction Proliferation and Terrorism, which was formed after this incident, stated that 'Given the high level of know-how needed ... [we] should be less concerned that terrorists will become biologists and far more concerned that biologists will become terrorists.' All up, it would take the FBI seven years to investigate the attacks. Initially they focused on Dr Steven Hatfill, a US bioweapons expert. At the time, there were theories that the attack was the work of a rogue CIA agent. Hatfill worked for a US company, Science Applications International Corporation, which had produced a report on a mail anthrax attack for the CIA in 1999. This was seen by the FBI as a blueprint for the 2001 attack.

There were also some other strange coincidences – like, for instance, the fact that Hatfill claimed to have been in the feared Selous Scouts, an army unit in Rhodesia that used biological and chemical weapons. He certainly had lived in Rhodesia for a number of years from 1978 and attended medical school there. In addition, around the time of the mailing of the contaminated letters, Hatfill was taking ciprofloxacin, the antibiotic used as prophylaxis against anthrax exposure (but also indicated for many other infections). Further, the letters to senators Daschle and Leahy had a fake return address '4th Grade, Greendale School, Franklin Park, NJ 08852'. Professor Don Foster, a professor of English at Vassar College, uncovered that the 'Greendale School' was the colloquial name for the Courtney Selous School in Rhodesia, in the suburb of Greendale, close to where Hatfill had lived. He also found an unpublished novel about a biological attack on Washington DC by Hatfill, and also that some of the hoax letters were sent from locations where Hatfill had been present. This started the search for an insider as the perpetrator, and others like Barbara Hatch Rosenberg, founder of Scientists Working Group on Biological and Chemical Weapons, were also convinced that the anthrax attack was the work of an insider, or even a CIA project that went wrong. Rosenberg published a document titled 'Possible

Portrait of the Anthrax Perpetrator', where she indirectly suggested Hatfill's involvement.

However incriminating the links to Hatfill may have appeared, the evidence was only circumstantial. Others claimed that Hatfill was being framed, and he himself protested his innocence. Eventually Hatfill won a $5.8 million lawsuit against the government and was exonerated. As the political and social pressure following the 9/11 and anthrax attacks increased, the FBI continued investigating the possibility of an insider attack. By 2006, the FBI shifted their focus to Bruce Ivins as the main suspect. A USAMRIID microbiologist at Fort Detrick in Maryland, Ivins had been hired to work on anthrax after the Sverdlovsk accident became public knowledge, making anthrax a top priority at the US Army's medical research facility. The anthrax spores Ivins worked with were cultured at Dugway Army Base in Utah. Aspects of Ivins' personality and background raised suspicions of his character. As a PhD student at the University of Cincinnati, he had developed an obsession with the Kappa Kappa Gamma (KKG) sorority after a student from the group spurned his romantic advances. In the decades to come he would continue his obsession with the KKG, breaking into chapters and stealing secret books from their collections. Later as a researcher at the University of North Carolina at Chapel Hill, he became obsessed with a PhD student, Nancy Haigwood, who was a KKG alumna, stalking her and vandalising her home and her fiancé's car. He also stalked two of his lab technicians at Fort Detrick and confessed to his psychiatrist that he had considered poisoning one of them after she left. It is telling of workplace culture and ethics that despite all of this, Ivins was nevertheless able to keep his job. As soon as the anthrax attacks occurred, Haigwood responded to a FBI call to American scientists to help identify the perpetrator, and named Ivins. Despite this, it would be several years before the FBI considered Ivins a suspect.

The circumstantial evidence against Ivins included the number of hours he spent in the lab. In August and September 2001 – the months leading up to the September attack – he clocked an abnormally high number of hours in the lab during the night and over weekends. Prior to the September 11 attacks, there was relatively lax security in laboratories. Whilst his hours were logged, he was able to work alone which gave him opportunity. The use of the Ames strain pointed to USAMRIID. The day before the first letters were postmarked, Ivins had taken four hours' leave – this was interpreted as possible time to drive to Princeton and back to post the letters at a mailbox just metres away from a KKG office. His colleagues argued that Ivins could not have prepared the anthrax without contaminating the lab and exposing other workers. In terms of motive, a possible one related to events around his livelihood, anthrax vaccines, leading up to the events of 9/11. In the late 90s, US troops were vaccinated against anthrax because of rumours that Saddam Hussein might use anthrax against them. However, following their vaccination, some soldiers reported fatigue and other non-specific symptoms. Their condition became known as 'gulf war syndrome' and it prompted debate about the safety of adsorbed anthrax vaccine. A number of reviews and research studies were conducted, which in the end confirmed the safety of the vaccine. Nonetheless in 2000 the US ceased vaccinating troops.

A messy investigation

In 2000 I had begun working at the National Centre for Immunisation Research in Australia, under the inaugural director, Professor Margaret Burgess. At the time, the anthrax vaccine was not registered in Australia, but one of the first tasks Margaret assigned to me was to prepare a safety briefing and fact sheet on it. Always ahead of the game and on top of every piece of available evidence around vaccines, Margaret knew this would be a contentious issue. The document I prepared concluded that the anthrax vaccine was safe. In 2002 the US Institute of Medicine conducted a review of the safety and effectiveness of the vaccine, concluding that while transient local and systemic reactions were common, it was safe and effective.

The FBI speculated that Ivins, gutted by the cessation of the military anthrax vaccination program and with his own second-generation anthrax vaccine in development, wanted to drum up interest in anthrax again. Having even been asked to move away from anthrax and work on other pathogens, it was speculated that Ivins may have been keen to reawaken enthusiasm for an anthrax vaccine.

According to a joint investigation by *ProPublica*, PBS program *Frontline*, and McClatchy, the case against Ivins was not that clear cut. The FBI, who themselves were novices in anthrax microbiology at the time of the attack, turned to USAMRIID for assistance. And herein lies the same conflict of interest that beset the SARS-CoV-2 origins investigation in the US: the lack of independent expertise by the investigating agency necessitates turning to experts, in this case virologists, who also stand to lose the most if proof of a lab leak were substantiated. That may explain why US intelligence agencies 'determined' by February 2020 that SARS-CoV-2 had a natural origin.

In the case of the 2001 Amerithrax attacks, the FBI soon realised the possibility of insider threat and were caught between a rock and a hard place. They set about testing the morphotypes (morphs) of all known Ames strain anthrax to find a match to the anthrax in the letters. The morphotype is a kind of signature based on how the anthrax grows on an agar plate. The FBI alleged that Ivins only supplied selected samples which did not contain the morphotypes seen in the anthrax letters. In 2002 he sent samples from his flask of anthrax that did not contain any morphs. When the FBI took their own samples from the same flask, they found a match.

The *ProPublica* investigation showed that Ivins made four anthrax samples available to the FBI between 2002 and 2004, and that some of these matched the samples in the letters. They reasoned that if Ivins were guilty, he would not have shared the samples. One of these was a sample that had initially been rejected because it was in the wrong kind of test tube. Others claimed that Ivins did not realise that subtle differences in morphs could point to the origin of the samples. Yet others said that by 2004, with the FBI swarming all over USAMRIID, Ivins had no choice in handing over samples. In the end, of all the Ames strains tested, including samples from overseas, only eight out of thousands matched the attack strains, and all came from USAMRIID, in fact from Ivins' flask.

Investigations, however, were complicated by the fact that in December 2001 and again in April 2002, Ivins had decontaminated his lab with bleach, in breach of standard safety procedures which required involvement of a safety officer. He had also been testing for anthrax spores in his office and lab. Only years later, however, did the FBI seriously begin focusing on Ivins. By that time, any evidence would have been long gone. The *ProPublica*, *Frontline* and McClatchy investigation suggests that the rigour of the science behind the forensic biology would not have stood up in court. Further, more advanced methods such as next generation sequencing, available from 2006 onward, were not used.

The FBI increased their psychological operations on Ivins, who was beginning to crack under the strain. Ivins had a personal collection of firearms and a shooting range at his home, and after expressing intent to cause harm with his guns, his weapons were confiscated. In July 2008 he committed suicide after finding out that the FBI was about to charge him. The following month, the US attorney general at the time, Jeffrey Taylor, made a public statement asserting that material from Ivins' flask of anthrax was conclusively the attack strain of anthrax. Meanwhile others, including Senator Leahy, speculated that a single perpetrator could not have masterminded the attacks alone, and there had to be more than one person involved.

Whether Ivins was guilty or not, there were elements that did suggest him as a key person of interest. Firstly, the strain of anthrax did likely originate from a flask that he had access to. And secondly, he had a joint patent for an anthrax vaccine, which may have been seen as a motive. The scientific community, however, rejected the notion that one of their own could have perpetrated the attacks. In 2011 a National

Academies of Sciences expert panel concluded that the scientific evidence was not conclusive enough to point to Ivins as the perpetrator. Yet all of the prevailing theories focused on the perpetrator being an insider American scientist, or more than one person, because of the high level of specialised knowledge required, the unique domestic strain of anthrax involved and limited access to it. Senator Leahy maintained this was not a lone wolf attack, and that more people were involved. To this day, the 2001 anthrax attack in the US remains one of the most high-profile unsolved cases.

There but for the grace of God go I

Police and law enforcement are familiar with the concept of insider threat. In that context, insider threat entails cases when, for example, a trusted insider steals and sells trade secrets for payment or profit. A real example of such a case is Anthony Levandowski, an engineer at Google, who left the company and took numerous files relating to their self-driving car and created his own company which was acquired by Uber. Google filed a suit against Uber and in the end, Levandowski was sentenced for theft of trade secrets. Why do people do such things? According to the FBI, possible motives include greed, financial need, revenge, problems at work, ideology, divided loyalty, adventure/thrill or blackmail. Espionage for foreign governments is another possible motive. When it comes to medical research, we tend to think of scientists and doctors as having impeccable character. But character cannot be imbued by the award of a PhD or medical degree. The same spectrum of human character traits exists among doctors and researcher scientists as in any other profession, and the very same motives that apply to corporate insider threat also apply to biological research and healthcare. Personal profit through patents, scientific discoveries, high-impact publications, research grants, funding and fame – these are all motives for scientists to pose insider threat.

In another notorious case, in January 2003, Dr Thomas Butler, a plague expert at Texas Tech University and former US Navy scientist, reported 30 vials of plague bacteria missing, possibly stolen. If the samples really were missing, Butler did the right thing by reporting, but in the climate of fear around the preceding anthrax attack, his action triggered a substantial FBI response. Some also suggest that prior to the incident, Butler had a dispute over his research grants with the university and was facing suspension of his clinical research. The report of stolen samples may have been in response to that dispute. Initially Butler claimed that the samples were missing or stolen and signed a statement without legal counsel. But later he said he might have destroyed or sterilised the samples in an autoclave. Butler alleges that the FBI tricked him into saying the samples were autoclaved.

To this date, the fate of the missing samples reported by Butler is unknown. What is clear is that he did not follow standard procedures and shipped samples of plague to Tanzania without the proper approvals. Butler also had contracts from drug companies that were not declared to the university and may have been facing repercussions from the university. The upshot of all of this is that his report of the missing vials triggered the bioterrorism response plan and an investigation, which resulted in his conviction on 47 of 69 charges including shipment of plague samples to Tanzania without the proper declarations or permits, and of receiving undeclared payments from pharmaceutical companies. Butler was, however, acquitted on the charges of losing the samples.

Butler's imprisonment was met by an outcry from the scientific community. Among those who expressed their concern was Dr Donald A. Henderson. A leader of smallpox eradication, Henderson felt that the government was setting an example to scare other scientists into following regulations around security sensitive pathogens. Numerous letters of support for Butler were issued, including one signed by the presidents of the National Academies of Science and the Institute of Medicine, which read:

> Although we are not in a position to make any claims about Dr. Butler's guilt or innocence, various members of the National Academy of Sciences and the Institute of Medicine have expressed concerns to us, which we share, about the potential impact of this prosecution on Dr. Butler personally, which is quite troubling, and about the potentially alarming effect on the infectious disease research community of which Dr. Butler is a part. We are particularly concerned about the impact that Dr. Butler's case may have on other scientists who may be discouraged from embarking upon or continuing crucial bioterrorism-related scientific research – thereby adversely affecting the nation's ability to fully utilize such research capabilities in preparing defenses against possible bioterrorist attacks.

As a researcher, Butler was a pioneer of oral rehydration solution for cholera and other diarrhoeal diseases. This and his other research achievements were cited in letters of support, and his contributions to medicine were considered by the sentencing judge when handing down a lesser sentence (two years rather than the 10 sought by federal prosecutors). A letter of support for Butler in the journal *Clinical Infectious Diseases* stated that 'He has a wife (to whom he's been married for 25 years), 4 children (one recently graduated from Stanford, another recently graduated from the Massachusetts Institute of Technology, one is in junior high school, and one is in elementary school), and no sources of income.' The bulk of the letters in support of Butler suggested that scientists, by simple virtue of being scientists, warrant special treatment compared to other people accused of crimes and should be allowed a lower standard of conduct, and that their freedoms should not be impinged upon. There was also the suggestion around some of the issues, such as the undeclared contracts and bending of rules around shipping of samples, that 'everyone does it'.

Espionage?

Throughout my career I have encountered examples of people not acting ethically, and in one case, I left a job because I could not, in good conscience, stand by knowing that something improper was occurring. Rather than leaving, another option would have been to be a whistleblower, but whistleblowing usually is a career-ending move. In the case of Butler, I believe that the scientific community stood by in horror, thinking 'there but for the grace of God go I'. The fear of less freedom to bend the rules may be what galvanised the scientific community. Most of the letters were written by fellow scientists, who would have been chilled by the degree of scrutiny Butler's actions were subjected to. This perspective is radically different from the perspective of law enforcement professionals I have spoken to about the case – they do not believe laws should be applied differently to some people, or that some people deserve more special treatment than others.

The disconnect between the scientist and law enforcement perspectives on insider threat is something that needs to be addressed to improve biosecurity. In the case of Butler, his career in science in the US was over, and he later worked in the Caribbean and Saudi Arabia. In 2010 Butler made the news again, this time for possessing a suspicious metal canister at Miami airport when he arrived from Saudi Arabia, where he had been teaching. The airport was evacuated, and Butler detained, but later released without charge after the canister was tested and found to be clean.

In a more recent case, in July 2019 Dr Xiangguo Qiu, a leading Ebola scientist at Canada's high security BSL4 National Microbiology Laboratory in Winnipeg, was evicted from the lab along with her husband, Keding Cheng, and several of their PhD students. Qiu was responsible for developing the Ebola treatment ZMapp, and had won many awards for it, working closely with Ebola expert Dr Gary Kobinger at the same lab. No one knew why Qiu and her husband had their security clearances revoked or were expelled from the lab, and the Canadian government has suppressed information. In the two years since Qiu's expulsion, it has been revealed that she had made multiple trips to China over the years, had close collaborations with the Wuhan Institute of Virology, and had shipped samples of Ebola and Henipah virus to that same lab. Several of her PhD students had come from Chinese institutions linked to the Chinese military, including one from the People's Liberation Army Academy of Military Medical Sciences.

Although the reason for Qiu's dismissal remains a mystery, it may have been because of suspected espionage, given she was named on patents filed in China, without any attribution to her employer, the Public Health Agency of Canada. Qiu claimed she did not know she had been named on a patent. The Canadian government may have asked why a salaried Canadian scientist should contribute intellectual property to inventions in another country, with no benefit to Canada. Dr Kobinger, now the director of the US National BSL4 Laboratory at Galveston, Texas, defended her, saying '[Qiu] told me, "This is a misunderstanding and I don't know why I was walked out of the building." She didn't understand. She was, from the bottom of her heart, saying that this is a misunderstanding.' Some reports suggest Qiu took shortcuts with paperwork, and a former colleague said, 'Scientists are a weird bunch. They're willing to take the [funding] … but don't like the idea of following all the rules.' This response resonates with the pushback by the scientific community over the prosecution of Butler and the fear of more widespread scrutiny. In 2021 it was reported that Qiu and her husband had relocated to China. We may never know the reasons behind these events, but they highlight that espionage is one possible reason for insider threats.

A police chief's insights

When I first started researching insider threats in biological sciences in 2015, there was not much to find about it except for articles on Bruce Ivins and Thomas Butler, mostly in support of them. Then I came across a paper by Thomas (Tom) Engells, titled 'The Insider Threat – A New Aspect of Biosecurity'. It was a compelling paper. I realised that the author was not a researcher or academic, but chief of police at the University of Texas Medical Branch and responsible for biosecurity at the BSL4 National Lab in Galveston – one of the few labs in the US that does Ebola research. Engells wrote that 'Several of the select agents are viruses, and it is natural that live viruses reproduce. Thus, accurately accounting for these substances becomes very dynamic and even more problematic to effectively secure. For unlike nuclear materials or chemical nerve agents, the biological select agents can reproduce and exist in great quantities unregulated in nature.'

At the time I encountered Tom Engells' paper, I was launching the course Bioterrorism and Health Intelligence for the first time at UNSW. It was to be part of a suite of courses offered through the PLuS Alliance at our partner university Arizona State, and so PLuS supported the design of the new course. We lined up a stellar, international list of speakers, including NSW Police and the FBI. In addition, I cold-called Engells, sending him a hard copy letter. Happily, he accepted. Tom became a regular in our course until his untimely and sudden death in January 2018. When I visited Tom Engells in Texas in January 2017, his wife, Peggy, took my kids out for the morning for a tour of Galveston, while he organised a senior virologist to give me a tour of the BSL4 lab. Tom's experience with the Galveston National Laboratory together with his law enforcement and military background as a former marine gave him a unique insight into biosecurity.

There is still a gap in the medical literature about insider threat in laboratories, and it took a law enforcement professional to write about it. I recall a conversation I had with Tom Engells about the disconnect between law enforcement and scientists that was so apparent in both the Thomas Butler and Bruce Ivins cases. He agreed with me that there needs to be a bridging of this gap. He agreed that a better understanding needs to be reached between law enforcement and science, and that vested interests of each profession need to be overcome for the greater good. In the end, if scientists are involved in lab accidents or, worse, deliberately nefarious conduct, it affects all scientists and the regulation of research in every laboratory. Both Tom and I shared the belief that that was why there was such vehement opposition to the investigation of Ivins and Butler.

In December 2017, I attended a WHO meeting in Geneva. The same virologist who had given me the tour of the Galveston lab was present at the meeting. I said hello to him and let him know that Tom Engells had just been in Australia to teach in our course at UNSW on insider threats and bioterrorism. I was taken aback by what the virologist said in response: bioterrorism and insider threats, he claimed, were 'all a load of nonsense'. This reaction neatly encapsulates the problem of denial by the scientist community, and why insider threats may not be detected in laboratories (until it is too late). However, man-made epidemics may arise simply by a lab accident, or, as we say, by error, not terror.

3

ERROR, NOT TERROR

LAB LEAKS AND MISHAPS ARE MORE COMMON THAN you may realise. They may involve a lab worker accidentally becoming infected and spreading the infection in the community. Or they may involve failed inactivation of a virus for vaccine production which results in thousands or even millions of people being injected with a deadly, live virus. Many vaccines, including the first ones for pertussis (whooping cough) and some for COVID-19, use a whole live pathogen that is killed in the lab. This can then be used as a vaccine without causing illness, as long as the inactivation (or killing) of the virus does not fail.

Sometimes, however, the inactivation process does not work as it should. Such a case occurred in the United States in 1955. It happened at Cutter Laboratories, which were then producing the Salk inactivated polio virus vaccine for the first mass vaccination program against polio. Due to inexperience and lack of expertise at the lab, the vaccine was not properly inactivated. This resulted in over 200 000 children being injected with live polio virus. As a consequence, about 200 of these children developed paralytic polio and 10 died.

This tragic event, which became known as the Cutter Polio Vaccine incident, had serious flow-on effects in public health management. The incident produced a loss of public trust in vaccines, setting back vaccination programs generally. Other labs at this time were successfully producing the inactivated polio vaccine, but it only takes one mistake to damage an entire vaccination program. The Cutter Polio Vaccine incident led to federal regulation of vaccines in the US. Today, any product being used must undergo regulatory approval and testing in the Food and Drug Administration.

Another kind of accident may involve pathogens being accidentally pumped out of a facility, as occurred in the Sverdlovsk anthrax accident in 1979, or in the Zhongmu Lanzhou biopharmaceutical plant brucellosis incident in 2019. Brucella is a bacteria recognised for potential use in bioterrorism. In nature, brucellosis is usually acquired by humans through contact with infected animals. The disease caused by brucellosis takes up to four weeks after exposure to develop and includes a flu-like illness which has a fatality rate of about 2 to 5 per cent. It can also result in abnormalities of the blood, immunological complications, congenital malformations, meningitis, endocarditis and foetal death. As a bioweapon, the most feared deployment of brucella is via airborne release.

In late 2019, while the SARS-CoV-2 pandemic was also beginning, an extraordinary lab leak occurred from the Zhongmu Lanzhou biopharmaceutical plant in the Gansu province of China, which was producing brucella vaccines for animals. Lab attendants were using expired disinfectant which meant that the bacteria were not killed as they should have been, resulting in live bacteria being present in the waste. Around

August 2019 waste in the form of an airborne plume leaked from the factory, releasing massive quantities of brucella into the surrounding region. For almost six months, the dangerous leak went on and on, with no intervention to stop it. For at least a year, from late 2019 to late 2020, cases continued to occur, and it wasn't until January 2020 that the factory had its license revoked. By that time over 10 000 human cases of brucellosis had occurred in the surrounding area. Early reports by the Chinese government, however, suggested numbers were much lower. The company, meanwhile, covered up the accident, but after 3000 initial infections, the media began to report on the outbreak, as more people kept getting infected.

The scale of the damage was immense. Environmental and livestock contamination was so extensive that the disease spread to farms as far east as Shaanxi province over 600 kilometres away and as far south as Inner Mongolia over 400 kilometres away, likely through infected farm animals being traded to those areas. The mortality rate for brucella when untreated is up to 5 per cent – higher than SARS-CoV-2 – often due to infection of the heart or brain. No deaths were reported, but no official data have been released, and there has been sparse media follow-up of the event. We also do not know the extent of public health alerts to the affected population following the incident, but it is possible there were anywhere between 200 and 500 deaths arising from over 10 000 infections.

The accident at the Zhongmu Lanzhou biopharmaceutical plant may have been bigger than the Sverdlovsk anthrax incident, but we may never know. It is likely there has been some cover-up of the incident and its true impact. All we can piece together, however, is that the government delayed taking action, with the result that further preventable infections were allowed to occur. As for would-be bioterrorists, the incident shows the devastating impact on humans and livestock of airborne brucella, and how the effects can spread far beyond the initial site of release.

Lab leaks

Major safety breaches with dangerous pathogens are common in leading institutes around the world. In 2014 a vial of smallpox was found lying around in a cupboard in the National Institutes of Health in the US. That same year, 75 staff at the US Centers for Disease Control and Prevention were accidentally exposed to anthrax, and also to Ebola in a separate incident. A better-known incident is the death of Janet Parker following exposure to lab-leaked smallpox in the UK in 1978. Parker, a photographer, happened to work on the floor below a laboratory run by Henry Bedson, a microbiologist who was experimenting with variola (the virus that causes smallpox). Parker herself had no direct contact with the lab, but the lab had sloppy practices and did not meet WHO standards for biosafety. In the investigation that followed, it was found that the virus had likely been transmitted by air currents through the service ducts from floor to floor. The international repercussions of the incident included the delaying of the certification of eradication of smallpox for another two years.

The American Biological Safety Association's catalogue provides a survey of global incidents of viruses and bacteria that have been the subject of lab leaks and accidents. Some of these incidents have had fatal consequences. In 2010, garden variety salmonella in a lab in the US caused one death and numerous infections. And in 2004 a lab scientist, Antonina Presnyakova, at the Vector Institute in Russia died after exposure to Ebola. Presnyakova had been engaged in inoculating guinea pigs with the virus and accidentally pricked herself with the needle. Fortunately for others around her, she was quarantined so did not infect anyone else. But tragically for her, it was a fatal accident.

Simple accidents like this can and do happen. Despite proper personal protective equipment (PPE), a needle can still pierce two layers of gloves. This is a lesson I've learnt through personal experience. In the 1980s I had an accident while working as a medical registrar at a Sydney teaching hospital. Called in at night to give an intravenous treatment to a patient who was known to be HIV positive, I used double gloves and a butterfly needle to provide the treatment. A butterfly needle sits at the end of a long piece of tubing, which is packaged with the tubing in a tight coil. As such, the tubing has the potential to curve and bend unpredictably. After injecting the treatment, I removed the butterfly, which was on the prep tray, and began peeling a Band-Aid for the patient. In retrospect, I should have disposed of the needle first, but the patient was bleeding from the injection site, so I wanted to apply the Band-Aid first. The butterfly tubing, which had the patient's blood in it, had twisted, resulting in the needle pointing straight up from the tray. As I peeled the Band-Aid, the needle pierced through both layers of gloves and the back of my left hand. Luckily for me I did not get infected. The patient was taking part in a trial of the first HIV treatment, zidovudine, which probably meant their viral load was low. I too was given a six-week course of zidovudine, which at the time had to be taken every four hours or so. I still recall the sound of my alarm going off through the night, alerting me to wake up and take my medication.

All it takes for a lab accident to occur is human error. And human error is exceedingly common. The implications for such accidents are potentially great. Lab accidents involving contagious pathogens may have impacts beyond the individual, and have the potential to spark an epidemic, as transpired in 1977 when a widespread pandemic of influenza H1N1 broke out. Referred to at the time as 'Russian flu', the virus was found to have an identical genetic sequence to a strain that was extinct and had not circulated for over 27 years. Today it is acknowledged that this pandemic was caused by a lab accident or escape of the virus from experiments on a live influenza vaccine being undertaken in Russia or China. But at the time, at the height of the Cold War, Western scientists went to great lengths to create a narrative that the virus had occurred naturally. Why did these scientists react in this way? Years later, analysis suggests that they did not want to offend their Russian or Chinese peers or pour fuel on the Cold War. And so they promoted theories that the virus had emerged after having been thawed from frozen form, or through exposure to a dormant strain of the virus in animals – the usual scapegoat. Yet multiple studies testing this theory found no evidence to support it.

It wasn't until 30 years later, in 2008, that experts began to concede that the 'Russian flu' pandemic was unnatural in origin. So-called because the strain was first detected in the Soviet Union, in fact the 'Russian flu' arose simultaneously in the Soviet Union and China. One possible explanation for its appearance is deliberate release. Another is a lab leak. In 2015 scientists Dr Michelle Rozo (now Director for Technology and National Security in the US) and Dr Gigi Gronvall published a paper after studying the epidemic, including the genome of the virus. In the paper, they argued that the most likely scenario was a lab leak from research to develop a live, attenuated influenza vaccine. The lab leak theory was supported by the fact that the epidemic virus had a signature genetic change for temperature sensitivity – something found only in engineered influenza vaccines. This means that the virus cannot replicate at higher temperatures and becomes less virulent as a result. This kind of manipulation is carried out when developing live, attenuated influenza vaccines. In fact, at the time, both China and the Soviet Union had been running trials of such vaccines in the years leading up to the epidemic. In addition, the live vaccine was being mass produced in Odessa (now in Ukraine). The evidence suggests that the vaccine was not properly attenuated and caused a pandemic and over 700 000 deaths.

The 1977 'Russian flu' affected young people in their 20s the most severely. In the UK and US, the epidemic affected military barracks, incapacitating young soldiers. In the US Air Force, 76 per cent of cadets became ill. Just like the 2009 H1N1 pandemic more than three decades later, older people were not badly affected because they had 'immune memory' to the virus due to past exposure to related or similar H1N1 strains. In their 2015 paper, Rozo and Gronvall discussed the deliberate release scenario, but concluded the lab leak theory is more credible, as influenza was not a high priority in the Soviet bioweapons program. They also lay out the various arguments of a natural origin and explain how these were politically influenced and changed over time, until finally, 30 years later there was general acceptance that the pandemic was the result of a lab leak. The authorities lining up behind the theory of natural origins at the time were impressive. WHO, the global arbitrator on such matters, for instance, denied any possibility of unnatural origin. It is interesting that at the time so many American scientists also vociferously denied a lab leak.

Coronavirus accidents

Lab accidents continue to occur on a regular basis, and sometimes these involve dangerous viruses. Scientists have been fascinated with the original SARS virus, which caused a deadly epidemic in 2003, and have continued to experiment with the virus. In 2004, four lab workers in Beijing, China, became infected with SARS in a lab that was apparently not working with live SARS coronavirus. The lab was working with inactivated SARS, so it is possible they did not properly inactivate the virus – or they were working with live virus despite the denials. SARS had a much higher fatality rate than SARS-CoV-2 – over 12 per cent compared to about 2 per cent respectively, so they were fortunate that none of the workers died. In 2015, SARS expert Professor Ralph Baric and his team at the University of North Carolina at Chapel Hill and his collaborator, Professor Shi Zhengli from the Wuhan Institute of Virology, also a leading SARS lab, jointly published a paper in *Nature*, where they described creating a chimeric SARS-like virus using reverse genetics.

This is a possible example of 'gain-of-function' research. Gain-of-function research confers unnatural, enhanced capabilities on a virus, a vexed issue that resulted in two bans, the first in 2012 and the second in 2014, in the US. Somehow this work was exempt from the ban on gain-of-function research in place at the time. Dangerous coronaviruses like SARS cannot ethically be tested on humans. Therefore, to study SARS and related viruses, Professor Baric developed humanised ACE2 transgenic mice, which are mice engineered to have the same human receptors for ACE2 that can bind SARS and SARS-CoV-2. He later sent these mice to the Wuhan Institute of Virology to help their coronavirus experiments, as the Wuhan lab had some interesting SARS-like viruses, but not the humanised mice in which to test them. These mice enabled gain-of-function research on coronaviruses in the Wuhan lab, which would later become the subject of intense scrutiny around a lab leak as the cause of the COVID-19 pandemic.

In 2016, a lab accident in Baric's coronavirus lab made the news. A scientist working at the lab was accidentally bitten by a mouse experimentally infected with a genetically engineered 'SARS-like' coronavirus. Public health authorities were not notified. Instead of quarantining the scientist, they allowed her to continue normal activities wearing a surgical mask and reporting her temperature twice a day. Fortunately, she did not get infected, but this was one of several near-miss incidents at the same university. The incident resulted in scrutiny from US federal authorities, including the National Institutes of Health and the Centers for Disease Control and Prevention, and freedom of information requests about the incident. The name of the specific coronavirus was redacted from publicly available material. Since that time, a further four breaches have occurred in the same lab, exposing up to six laboratory workers to engineered SARS-like coronaviruses, and two further workers were exposed to an engineered MERS-like coronavirus. Despite their exposure, all were allowed to carry on their lives as normal, potentially exposing others in the community. In April 2020, another mouse bite occurred, this time from a mouse infected with an engineered version of the SARS-CoV-2 virus. In this case, the health authorities were notified, and the laboratory worker was quarantined for 14 days.

In November 2021, a lab accident occurred with SARS-CoV-2 in Taiwan. A lab worker handling an animal deliberately infected for an experiment became infected with the identical strain of SARS-CoV-2. It is unclear how she got infected, but it is thought that she may have inhaled the virus after incorrectly removing her personal protective equipment after the experiment. The worker resigned, and her supervisor, who himself had been infected by the first SARS virus in the lab in 2003, also retired. The incident was disclosed publicly by the Taiwanese government, and fortunately did not result in further spread. However, following investigations, many safety breaches were discovered at the lab.

It is easy to see how a lab accident can result in an epidemic or even pandemic. After all, a pandemic is an epidemic that spreads globally – and for that to happen, it needs to be a pathogen that is highly contagious. Once the genie is out of the bottle (or out of the lab), the virus takes care of the rest of the work by spreading. The examples above reflect the various problems associated with the response to such events – from disclosure to attempts to manage incidents internally, and the refusal to reveal details until decades after the event. There are many influences at play, including political forces and vested interests.

If the SARS-CoV-2 outbreak, which has wreaked such global havoc over the past few years, proves to have been the result of a lab leak, past experience (including the 1977 influenza epidemic and the Sverdlovsk anthrax accident) indicates that we won't be told for a very long time. For the time being, we are informed that it is a coincidence that SARS-CoV-2 arose in Wuhan, a short distance from a high security lab, the Wuhan Institute of Virology, well known for coronavirus research. This research included investigations into the closest known bat virus relative of SARS-CoV-2, RaTG13, which came from a bat cave thousands of kilometres away in Yunnan province. Even the very existence of RaTG13 wasn't disclosed until after the pandemic had begun. We are also told to ignore the fact that half of the first recognised cluster of cases did not visit the wet market in Wuhan. And we are also directed to overlook several other lines of evidence indicating that the epidemic in fact started earlier than December 2019, with lab workers at the institute reportedly falling ill in November. The US intelligence reports based on satellite and signals data also suggested a major incident at the Wuhan Institute of Virology as early as October 2019. But that, too, we are told to ignore. Instead, there has been a veritable industry in opinion pieces, counter-narratives, publications and selective research loudly shouting down lab leak theories and insisting that COVID-19 arose from nature.

Even antibody studies from Europe and the US show the virus was already circulating in both places by December 2019, which means that it must have emerged prior to that time. On 31 January 2020, emails between top virologists, released under a freedom of information request, showed that initially, most experts thought that the SARS-CoV-2 outbreak was the product of a lab leak. Dr Alina Chan and Dr Matt Ridley, co-authors of *Viral: The Search for the Origin of Covid-19*, cover this in their book. They point out that evolutionary virologist Professor Edward Holmes from the University of Sydney was '80 per cent sure this thing had come out of a lab'. Professor Kristian Andersen from the Scripps Research Institute in California thought it was 60 to 70 per cent likely to be a lab leak, and most others agreed. Sir Jeremy Farrar, director of the Wellcome Trust, pegged it as 50 per cent likely to be a lab leak. Following an emergency meeting held between these virologists and professors Anthony Fauci and Francis Collins of the US National Institutes of Health (which had funded some of the research at the Wuhan Institute of Virology), the group changed their position. In April 2020 they quickly published a paper in *Nature Medicine* which said, 'we do not believe that any type of laboratory-based scenario is plausible'.

As time goes on and each piece of evidence supporting a lab leak is exposed, a new paper is published in a scientific journal pushing back and promoting the natural origins theory. This tug of war between narratives and counter-narratives is a form of information warfare, which I will explore later in this book.

The lab next door

It's not just official, highly regulated laboratories that can suffer lab leaks and accidents. Unregulated or citizen science labs are also prime locations for such incidents to take place. DIY biology is the latest global craze. Citizen science labs like BioFoundry in Sydney and BioQusitive in Melbourne have been operating for well over six years now. And there are similar labs in most major cities in the world. Sydney's BioFoundry was started by Meow-Ludo Disco Gamma Meow Meow, best known for implanting his Opal public transport card chip into his hand. The philosophy behind citizen science, variously known as biohacking (which generally refers to manipulation of genetic material) or DIY biology, is that scientific inquiry should be free to all. However, DIY biology is poorly regulated, guided by 'codes of conduct' and has minimal ethical oversight.

Certain types of research are required to get ethical approval, but there is no system of enforcement in DIY labs and no means of ensuring that such research actually receives it. Research conducted in universities has to undergo various safety and ethical reviews and gain approvals before it can proceed. Usually this involves submission of an application for approval, and an iterative process (which may take months) of addressing concerns and revising the proposal to obtain final approval. Research not approved in universities could be conducted in DIY labs, which are more difficult to monitor in the community. Biohacker labs are typically given Physical Containment Level 1 (PC1) or BSL1 levels, which is for the working on microorganisms that are thought to pose a low hazard. In Australia, biohackers planning to work with genetically modified organisms (GMOs) are supposed to apply to the Office of the Gene Technology Regulator for a licence and must not release any GMOs into the environment. The problem is that this all operates according to an honour system, and there is no policing of what actually happens. What happens to the biological waste produced at DIY labs? We just don't know, and there are many unfortunate possibilities, including dumping in the community, at universities (by biohackers who also work in universities), or elsewhere. Further, whilst public biohacker labs are known, there is no way to keep track of covert labs in the community.

Just as clandestine drug labs have proliferated in the community, so too may DIY biolabs. These days it's not that hard to procure the basic equipment. A lab-in-a-box kit can be purchased online, as can most of the equipment and technology required, including genetic code. If lab leaks are common in top notch research labs, then we can only assume that a lab leak from biohacker or illicit labs will be even more common, and, what's more, far harder to track. As for leaks, the greatest concern is enhanced or 'super' pathogens that have been engineered or created synthetically in a lab, for they may have the potential to cause a pandemic.

4

DR JEKYLL AND MR HYDE OF BIOLOGICAL RESEARCH

IN 2001, AUSTRALIAN SCIENTISTS RON JACKSON AND Ian Ramshaw engineered a cousin of smallpox, one of the most deadly diseases known to humanity, and unexpectedly found they had made it vaccine resistant *and* universally fatal. Mousepox is caused by an orthopoxvirus, ectromelia virus, which is closely related to monkeypox and smallpox, but only infects mice (typically causing a mild illness in them). Jackson and Ramshaw were trying to develop a contraceptive vaccine for mice to help control their breeding and mitigate agricultural damage. They used a virus called ectromelia (or mousepox) as a vector to carry mouse egg proteins so that the mice would develop antibodies to their own eggs and thereby become infertile. They inserted a gene for production of interleukin-4, a cytokine that stimulates the immune system, hoping this would accelerate the production of antibodies. Instead, it killed all the mice. When they tried to vaccinate the mice with smallpox vaccine (which also protects against ectromelia virus), they discovered their genetic engineering also made the mice resistant to smallpox vaccine. In effect, they had created a double whammy: a more deadly virus that was at once also vaccine resistant.

Jackson and Ramshaw's work on mousepox sparked concern around the world. It suggested that similar engineering of smallpox had the potential to produce a more lethal version of the virus which current vaccines would not be able to protect against. Natural smallpox has a 30 per cent fatality rate. The mousepox experiment, by contrast, killed all the mice – what if a similarly engineered smallpox resulted in a fatality rate of close to 100 per cent? If current vaccines could not protect against such a virus, then a pandemic of engineered smallpox could be an extinction event. The situation triggered new research on countermeasures to protect against weaponised smallpox, including development of new antiviral drugs.

In conducting experiments with mousepox, Jackson and Ramshaw were trying to benefit humanity by protecting Australian agriculture from the pestilence caused by mice. In the end, however, their research had an unexpected side effect, causing widespread concern. It constitutes an example of what is known as dual-use research of concern, which refers to research or technology intended to benefit humanity which may also be used to cause harm. It is effectively like an evil twin, so to speak. The concept of dual-use research of concern applies to many technologies – among them, biology, artificial intelligence, cyber technology, nanotechnology, drones or robotics. Some of these technologies, despite their valuable uses, pose existential threats to human survival.

Gain-of-function research

The top-ranked threats that may cause human extinction include natural disasters (such as an asteroid hitting the earth), a nuclear holocaust, a pandemic, climate change and rogue AI. If, for example, machines were to become more intelligent than humans, then the human race may face extinction. Equally, if a deadly new pandemic with high mortality rates emerges, it may be an extinction event or cause mass depopulation. In the case of COVID-19, it has caused a bare minimum of 15 million deaths globally in two years and has reduced life expectancy in many countries. In the US, for example, COVID-19 has produced a drop in life expectancy of two years. In the UK, meanwhile, in early 2022 the coronavirus was causing hundreds of deaths a day and thousands a week – the equivalent of a Boeing 747 plane crashing every two days. Whilst not an extinction event, COVID does have the potential to cause significant depopulation of the world as it continues on its relentless path in coming years. An engineered pathogen – for example, vaccine-resistant smallpox – has the potential to cause even greater numbers of deaths.

In 2011 professors Ron Fouchier (Erasmus University in the Netherlands) and Yoshihiro Kawaoka (University of Wisconsin-Madison and the University of Tokyo) sought to publish the results of their gain-of-function research. Gain-of-function research refers to a specific type of dual-use research of concern. It entails engineering pathogens in such a way that they gain properties which they don't naturally possess, and which make them more dangerous to humans by being more transmissible or causing greater severity of disease. This process can be done by splice and dice genetics (deleting or inserting specific genes) or by repeatedly transmitting a virus through an animal species until the virus adapts to that species and becomes contagious for it. This is called 'serial passaging' and does not leave tell-tale signs of genetic engineering. Unlike splice and dice genetics, the results look natural, and the resulting virus will appear as though it has been around for years, decades or even centuries, when in truth its evolution has been accelerated in a lab.

In the case of Fouchier and Kawaoka, their research involved the engineering of an avian influenza virus not naturally contagious between humans so that it was made easily transmissible (thereby transforming it into a virus with pandemic potential). Their research opened the floodgates to the new age of engineered viruses. At that time, no countries outside of the United States even had mechanisms for assessing potentially dangerous research like this. The US had set up the National Science Advisory Board for Biosecurity (NSABB) in the aftermath of the 2001 anthrax attacks. These attacks, which became known as the Amerithrax attack, led to the publication of the Fink Report in 2004. Along with advising the creation of the NSABB, other key recommendations of the Fink Report included review of plans for risky experiments and of publication of such methods.

In December 2011, the NSABB recommended censorship of publication of the full research methods employed by Fouchier and Kawaoka in the journal *Science and Nature*. At that point, there was a voluntary moratorium on engineered H5N1 research, but it was then followed by a subsequent furore among virologists and scientists, who began lobbying in earnest for gain-of-function research. The issue divided the medical and scientific community. Virologists and those with a vested interest in freedom of gain-of-function research claimed that science should not be censored, and scientists should be free to pursue their research. Over 1000 scientists signed a petition supporting full publication of Fouchier and Kawaoka's research into avian influenza, stating that

> Only [by releasing the genetic sequences] can researchers establish and track the global pattern of the evolution of the bird-flu virus. And, hopefully, work quickly toward an effective vaccine should H5N1 develop into a pandemic ... The future of humanity may very well be at stake here – there is no longer any time to waste.

Another group of experts, including Professor Michael Osterholm from the University of Minnesota (who was also a NSABB member at the time) and the late Dr Donald Henderson, who led the global smallpox eradication effort, rang the warning bell. They pointed out that gain-of-function research could be used for nefarious purposes or be subject to lab leaks. Writing in the journal *Science*, Osterholm and Henderson argued that

> disseminating the entirety of the methods and results of the two H5N1 studies in the general scientific literature will not materially increase our ability to protect the public's health from a future H5N1 pandemic. Even targeting dissemination of the information to scientists who request it will likely

not enhance the public's health. Rather, making every effort to ensure that this information does not easily fall into the hands of those who might use it for nefarious purposes or that a biosafety accident resulting in an unintended release does not occur should be our first and highest priority.

After intense lobbying by scientists, in 2012 the National Science Advisory Board for Biosecurity voted to allow publication of two papers by Fouchier and Kawaoka in *Science and Nature*. It was a momentous decision. Since then, a multitude of other gain-of-function studies have been published, including engineering of other avian influenza viruses, a SARS-like coronavirus and studies that identified genetic markers of transmissibility of various viruses in mammals and humans. A plethora of gain-of-function methods for creating mutated avian influenza and coronaviruses which can be readily transmitted between humans were published and available. The NSABB itself was essentially knee-capped following its decision, with loss of funding, restriction of remit and a cleanse-and-replace of the membership. However, the unrest continued. In 2014, there was another pause on gain of function. The following year, the US National Institutes of Health contracted Gryphon Scientific to conduct a risk analysis on paused gain of function research. The US and the European Union both released reports on the risks and benefits of gain of function during that time, but both were vague and inconclusive. This was a long overdue recognition of the risk, but critics have raised conflicts of interest in the process and a lack of global governance or risk assessment. Despite the pause, some apparent gain-of-function research, including on coronaviruses, continued.

Origin of SARS-CoV-2

During the moratorium on gain-of-function research following Fouchier and Kawaoka's research, the US-based EcoHealth Alliance provided funding from a US government grant to the Wuhan Institute of Virology for coronavirus research. During a Senate hearing in January 2022 a fiery exchange erupted between Senator Rand Paul and Dr Anthony Fauci, director of the National Institute of Allergy and Infectious Diseases. Paul accused Fauci of funding gain-of-function research in Wuhan, which Fauci denied. In an interview with *Scientific American* published online in March 2020, Professor Shi Zhengli from the institute stated that when she was notified about the emerging coronavirus epidemic, she wondered 'Could [the epidemic] have come from our lab?'

The Wuhan Institute of Virology is the world's leading bat coronavirus research lab and home to over 15 000 bat virus samples, as well as live bats. Professor Ralph Baric at the University of North Carolina at Chapel Hill first began working with Professor Shi in 2013. Shi had discovered the bat virus SHC014 but had trouble culturing it in her lab, and Baric approached her about a collaboration that would enable him to get SHC014 into his lab and use reverse genetics to engineer a mutant coronavirus with the features of the original SARS virus, which he possessed in his lab. Together, Baric and Shi published a paper in *Nature* in 2015, showing how reverse genetics could be used to engineer a chimera of SARS and Shi's bat coronavirus, SHC014, that was able to replicate efficiently in the human respiratory tract.

As noted in the previous chapter, Baric was a pioneer in the use of humanised mice for coronavirus research. Known as ACE2 transgenic mice, these creatures are engineered to carry human ACE2 receptors (the receptors that bind SARS-CoV-2). Since 2007 multiple scientists have used such humanised mice for coronavirus research. It is possible to take a sample of coronavirus from bats that is not highly contagious between humans and using serial passaging through humanised mice make it contagious in humans. The Wuhan Institute of Virology itself did not have access to such mice, but in 2016 Baric obtained approvals to export his humanised mice to the institute, confirmed by Professor Shi Zhengli. Some dispute that Baric's research and the research at the institute was gain of function, as it was all carried out during the moratorium, and was not allowed as a condition of the EcoHealth Alliance grant. The connection between the lab leak theory, US researchers and grants might explain why there was such a pushback against lab leak theories by the US scientific community and funding agencies.

The first WHO mission to investigate the origins of COVID-19 was criticised on various fronts: for allegedly being biased, for succumbing to political pressures, and for including members in its investigating team who had clear conflicts of interest, among them the head of EcoHealth Alliance, which funded some of the bat coronavirus research at the Wuhan Institute of Virology. Being heavily comprised of virologists, the team would have had a vested interest in the outcome. If the pandemic were determined to have been the result of a lab leak, its catastrophic impact would make such a finding impossible to ignore. It would have significant implications for virology, likely resulting in changes to oversight and lab practice globally. After continued criticism of the initial WHO investigation, a new committee was established. In late 2021, the Scientific Advisory Group for the Origins of Novel Pathogens (SAGO) was tasked with investigating the origins of new pathogens. Again, the composition of this group was over-represented by laboratory scientists.

There has been much discussion about the cause of the COVID-19 pandemic, about whether it emerged in nature or because of a laboratory accident. A group of scientists, some of whom were collaborators with the Wuhan Institute of Virology, issued an early statement declaring the pandemic to have emerged without any human involvement. Anyone who questioned this was labelled a conspiracy theorist. Very early in 2020, as outlined by Sharri Markson in her book *What Really Happened in Wuhan*, in record time, the US Intelligence Services also stated the pandemic was natural in origin. Markson made the point that tribalism was a powerful force in this. US President Donald Trump commenting on the lab leak theory, created a knee-jerk reaction by those who opposed him politically. It was not until 2021, with President Joe Biden in power, that the serious possibility of a laboratory accident was considered. This was a vitally important development. We can never hope to discover the truth about anything unless we can step out of our tribal alliances and examine all information objectively.

A question of probability

My own research has shown that because of dual-use research of concern and lab leaks, the probability of an unnatural pandemic is far greater than a natural one. To establish this, I used a simple epidemiologic risk analysis method. If there is a higher risk of a natural pandemic than an unnatural one, this would favour gain-of-function research. But if the risk of an unnatural pandemic is higher, this would caution against gain-of-function research. Using avian influenza as an example, I calculated the risk of a natural pandemic arising. This is a scenario I use as a teaching exercise for students in the Bioterrorism and Health Intelligence course. H5N1 is a common avian influenza virus. It mostly infects birds, and only occasionally infects humans who have close contact with sick birds. Because the virus receptors in the respiratory tract of birds are different from those in humans, bird flu does not spread easily between humans.

In my research I first set about quantifying the risks by estimating the probability of a natural pandemic occurring. I then compared these to the probability of an unnatural pandemic occurring as the result of a laboratory accident or deliberate release. Generally speaking, a pandemic may occur naturally (with a probability P1) or unnaturally because of a lab leak or biological attack (with a probability P2). Historically, natural pandemics occur every 10 to 40 years. This equates to a P1 of between 1/10 (10 per cent) or 1/40 (2.5 per cent). The probability of an engineered pandemic strain, meanwhile, is 1 (100 per cent). This is because such viruses have already been created in many labs around the world.

So therefore, the probability of a pandemic arising as a result of dual-use research of concern is calculated as follows: $P2 = 1 \times K$, where K represents either the probability of bioterrorism or a laboratory accident resulting in a pandemic. Working out the probability in this equation is more complicated, but given the high frequency of accidents in labs, the probability of a laboratory accident is high. The probability of a pandemic being the product of an act of bioterrorism or biowarfare is also likely to be high because these methods have been used throughout human history. Whenever new means become available and are disseminated, the chance immediately presents itself that these will be used. If, for instance, someone were to publish a method for making a 3-D printer gun on the internet, the probability of it being used would be 100 per cent. Similarly, while contemporary terrorist groups are making threats of biological attacks, if methods for genetic engineering of viruses were published, it is highly likely that these would be used. Using this reasoning, I estimated the probability of an unnatural pandemic was far higher than a natural one. There are many approaches that can be used to measure P1 vs P2, but the risk of unnatural pandemics is likely to remain much higher than natural ones. I use this risk analysis as an exercise in the course Bioterrorism and Health Intelligence, making students get into small groups, gather data and work out P1 and P2, and they all work out for themselves that unnatural pandemics are a greater risk.

Harm to humans as a result of dual-use research of concern may occur via two mechanisms: a laboratory accident, or deliberate release, and in the case of biology, it may result in an epidemic or pandemic. Dual-use research of concern due to genetic engineering has been documented as far back as the 1970s. In that period, the Soviet bioweapons program tried to create hybrid viruses from smallpox and Lassa virus, and to make pathogens more deadly and contagious. Genetic engineering has been greatly enabled by what's known as CRISPR-Cas9, a breakthrough in gene editing technology. It was pioneered by professors Jennifer Doudna in the US and Emmanuelle Charpentier in France, who won the 2020 Nobel Prize for Chemistry for their ground-breaking work. CRISPR-Cas9 has revolutionised medicine, agriculture, biosecurity and many other fields in which it has led to great gains for humanity. But it has equally enabled research that can harm humanity. The advent of CRISPR-Cas9 means that the clumsy, early attempts of the Soviet Union to create lethal chimeric viruses would be much easier and more successful were they to be developed today. If back in the 1970s the Soviets were trying to create super mutants as offensive weapons, then it is certain others will be doing so today. Other developments relevant to biosecurity include the emergence of 3-D printing of biological materials, an explosion of citizen science or DIY biology labs, and insect drones. These are robotic insects or real insects controlled through a radio-transmitter – a biological attack could utilise this technology for precision targeting of a biothreat.

Synthetic biology is another kind of dual-use research of concern. This entails the creation of a living organism in a lab from scratch. In 2002 scientists in New York created the first synthetic virus, a polio virus. Since then, many other viruses have followed, and hundreds of unregulated private companies have entered the business. Because any pathogen can be created in a lab, it is now no longer possible to eradicate a virus. Thanks to synthetic biology, for instance, the spectre of smallpox is now greater than ever. The only known repositories of the smallpox virus, variola, are the Centers for Disease Control and Prevention in the United States and the Vector Institute in Russia. Created in Novosibirsk in 1974, Vector was once the jewel in the crown of the Soviet bioweapons program. What became of the bioweapons held at Vector after the fall of the Soviet Union remains a mystery. For decades, there have been calls to destroy the variola stocks in Russia and the US, but despite plans to go ahead, the virus still exists.

In September 2019, the Vector lab in Novosibirsk blew up. The cause was allegedly a gas bottle explosion on the fifth floor, in a decontamination room that was apparently being renovated. The explosion was so severe that it shattered all the glass in the multi-storey building. The Russian government assured the world that no pathogens were stored on the fifth floor. Virologists around the world, despite having no apparent inside knowledge or training in the physics of explosions, immediately agreed with the Russian government and minimised the incident, saying the fire associated with the explosion would have killed any viruses.

No community voice

Over the years, occasional controversies have emerged around smallpox, as in 2001 when Australian researchers created a vaccine-resistant, lethal mousepox virus which is a member of the orthopoxvirus family and closely related to smallpox. The research raised concerns that it was a blueprint for engineering a more lethal and vaccine-resistant variola virus. With smallpox declared eradicated in 1980, since then, intelligence analysts have assumed the only way the virus would re-emerge in the world would be as a result of a lab accident or a break-in at the high security BSL4 labs at the Centers for Disease Control and Prevention or the Vector Institute – a near-impossible task. Given the hurdles that would need to be cleared to access lab samples, most intelligence agencies have assumed the threat of re-emergence of smallpox to be low. The possibility of smallpox being created synthetically had been dismissed as too technically difficult. Yet the greatly altered risk landscape that has emerged over the past two decades, attributable to advances in science, warrants a review of the risk of a re-emergence of smallpox.

Ethical questions around dual-use research of concern are still unanswered. There is a need for a global level of governance and accountability on these issues. When decisions made in one country can impact the rest of the world, what rules of conduct should underpin the creation of dangerous viruses? The normal processes governing medical research are not adequate to deal with these questions. Standard medical research is subject to regulation and approvals by Animal or Human Research Ethics Committees, but they only focus on individual risk – for example, the risk posed to a volunteer in a drug trial. Most gain-of-function research is done in animals, so typically, an Animal Research Ethics Committee will not consider harm to humans through contagion. Informed consent is fundamental to research – anyone who may be adversely affected by the research must be informed and must consent to it. In the case of a lab leak that causes a pandemic, informed consent is clearly absent for the most important stakeholder of all: the public.

As stakeholders, consumers are the most important. This is because the outcomes of dual-use research of concern, whether lifesaving drugs or a deadly pandemic, impact them the most. But sadly, the debate has been confined to scientists and has not been inclusive or comprehensive. The overriding concern in these discussions has been the freedom of scientists to conduct their research, and the rights of scientists have taken precedence over that of the public interest. What is lacking is a transparent process to address conflicts of interest in dual-use research of concern. In 2017, I conducted research to ascertain how much the general public knew about dual-use research of concern in Australia, the US and the UK. I found that 77 per cent of the people surveyed were entirely unaware of it. When provided with information about it, 64 per cent found it unacceptable or were unsure. Initially these people perceived the risk of laboratory accidents and deliberate bioterrorism to be low, but once they were provided with information, that figure rose. Given that they are the main stakeholders in the risks and benefits of dual-use research of concern, it is imperative that the community be part of the process.

5

JURASSIC PARK
FOR VIRUSES

IN THE BOOK AND 1993 FILM *JURASSIC PARK*, DINOSAURS were brought back from extinction, artificially created by humans in a lab. Because of the relative size and complexity of the respective genomes, creating a large, complex life form like a dinosaur is more challenging than creating a virus. Animal cloning first started with frogs in 1958 but became famous decades later with the creation of Dolly the Sheep in 1996. Animal cloning involves conferring a surrogate egg with the desired genes. The technique opens up the possibility of bringing creatures like the thylacine (or Tasmanian tiger) back from extinction. The thylacine was hunted to extinction by 1936 with bounties placed of £1 per head for dead adult thylacines and ten shillings for pups. Since 1999, with retrospective regret, scientists have been trying to clone the Tasmanian tiger. Horses, camels, livestock and pet dogs and cats have all been commercially cloned. Anything with DNA is now fair game: plants, animals and viruses. Even human beings.

In 2018 Chinese scientist He Jiankui caused an uproar by creating the first genetically engineered babies. Scientists in the UK and Sweden had been experimenting with editing human embryos since 2016, but He Jiankui was the first to bring such embryos to term, resulting in live births. If you thought the plot of the 1997 movie *Gattaca* was pure science fiction, you should be aware that it is very much a reality. The technology began in 2016 with the world's first three-parent baby. It will only be a matter of time before parents can routinely select genetic characteristics of their children.

Compared to creating humans, creating viruses is easy. The first synthetic virus, a polio virus, was created by researchers at Stony Brook University in New York, with funding from the Defense Advanced Research Projects Agency (DARPA). When their research first appeared in 2002, it created a storm of publicity about the danger of manufacturing viruses and the implications for biological terrorism. More synthetic viruses were soon made. In 2005, researchers at the US Centers for Disease Control and Prevention (CDC) used reverse genetics to recreate the influenza H1N1 virus that caused the 1918 pandemic and killed up to 50 million people. Like the researchers at Stony Brook University, the US CDC researchers published their findings in the journal *Science*. In order to understand the potency of the synthetic 1918 H1N1 virus, they infected mice with the disease and compared its effects with that of mice infected by a seasonal human H1N1 influenza virus. What they found was that the engineered 1918 virus was at least 100 times more severe than the seasonal H1N1. The mice infected with the engineered strain died within three days. In contrast, all the mice infected with the seasonal strain survived. They also shared the genetic make-up of the 1918 virus and laid out how it differed from the milder human strains.

Some of the scientific research community reacted with concern. Donald Henderson, the epidemiologist at the helm of smallpox eradication, with Professor Michael Osterholm who was on the NSABB at the time, warned of the dangers of such research in the event of a lab leak, to no avail. In 2014, Yoshihiro Kawaoka built on the 1918 H1N1 virus research to assemble the genes of modern avian influenza viruses and engineer a virus that closely resembles the gene sequence identified in the 2005 study. Kawaoka then tested multiple variations of the virus on ferrets (one of the best proxies for human influenza because they have the same receptors in the respiratory tract) until he found one that would spread easily between humans. In other words, he created a strain with 1918 pandemic potential.

In February 2022, a brand new, highly virulent variant of HIV, called 'VB', was discovered in the Netherlands. The location of this virus on the evolutionary tree suggests it arose in the 1980s or 1990s – this is perplexing given that HIV surveillance and research has been extensive and comprehensive for the past 40 years and should have uncovered it long ago. Another explanation for its sudden appearance and apparent age is gain-of-function research. As we've seen, gain-of-function research allows natural evolution to be accelerated by means of serial passaging through an animal host, so that decades of natural evolution can be replicated in weeks or months in a lab. What makes analysis difficult is that a virus resulting from gain-of-function research cannot easily be distinguished from a naturally evolved virus. The first undergoes the same evolutionary process as the second, simply at a much higher speed. Because the virus's evolution has been forced at high speed in a lab, if we ignore the possibility of gain-of-function research, it would look like it had evolved a very long time ago. In my view, in the case of the new VB HIV strain, the phylogenetic data showing it to be 40 years old suggest that it may well be the result of gain-of-function research. The possibility of man-made viruses is regularly and aggressively shouted down. And yet, the truth is, the technology enabling a sort of Jurassic Park of viruses has been available for 20 years. It would be simply naïve to think that such technology has not been used. And it would be foolish not to reflect on the motives for doing so.

SARS and MERS-CoV

Scientists have been fascinated with the SARS-1 virus since 2003 and have continued to research it in labs. The MERS coronavirus (MERS-CoV) has similarly attracted a lot of scientific interest. When MERS-CoV emerged in 2012, it had a much higher fatality rate than SARS. While SARS has a mortality rate of about 12 per cent, MERS-CoV is a whopping 36 per cent, which puts it in the same ballpark as smallpox. But MERS-CoV, which arose in the Arabian Peninsula and is largely localised to that region, is not highly contagious. Despite spreading to over 26 countries in the world through travel, over 80 per cent of cases have occurred in the Middle East. It never seeded epidemics in other countries, except for

South Korea, where a large hospital outbreak of 186 cases occurred in 2015. It sometimes caused hospital outbreaks in Saudi Arabia and surrounding countries, but most cases have been sporadic. And despite theories of camel to human transmission, over 60 per cent of all cases have no record of camel or other animal contact.

I myself became puzzled and intrigued by MERS-CoV after it emerged in 2012 and began researching it and applying epidemiologic principles to analyse its spread. Before the SARS-CoV-2 pandemic, my research program had published six papers on MERS-CoV, all investigating the unusual epidemiologic features of the outbreak. In 2014 we pointed out that overall, it did not seem very contagious, with a reproductive number (R0) less than 1. This means one case gives rise to less than one other case on average, which does not favour an epidemic. SARS-CoV-2 in comparison has an R0 of about 10 currently, meaning each case gives rise to 10 more cases and favourable epidemic conditions. The outbreak had grumbled along with mostly sporadic cases for ten years, and was geographically concentrated in the Arabian Peninsula, mainly the Kingdom of Saudi Arabia. In contrast to SARS-CoV-2, the pattern of MERS-CoV was unpredictable, with no seasonality, and variations in the pattern each year. Yet there were also some notable hospital epidemics where the R0 exceeded 1.

In 2014 I published a very controversial paper. In it I showed that, based on the epidemiological pattern and using a risk analysis approach, you could equally explain MERS-CoV as repeated deliberate release or by repeated infection of humans from animals. I did not claim it was one or another, simply that there were two possible explanations for the strange epidemiologic pattern of this virus, using principles similar to the Grunow-Finke Tool (GFT) – the most widely used epidemiologic tool for differentiating natural and unnatural epidemics. My team later published a paper where we applied the GFT to MERS-CoV, and it again showed a reasonable probability of this being unnatural in origin.

As we have already seen, health experts are generally uncomfortable with any suggestion that an outbreak could possibly be unnatural. As a result, I was aggressively attacked for my research. A whole gang of virologists published opinion pieces and attacked me on social media, refuting any suggestion that MERS-CoV was anything other than natural. This of course may well be true. And we clearly stated as much in our research. At the same time, others modelled the MERS-CoV epidemic to show it could only be explained by vast amounts of asymptomatic infection. This was categorically refuted by the antibody studies that followed – there was no evidence of widespread asymptomatic infection. Yet the clear message being sent to me was that it was not valid to even consider the origin or use recognised epidemiologic tools to do so. The Borden Institute of the US Army publishes a textbook on Medical Aspects of Biological Warfare which lists the GFT as an important risk analysis tool for assessing origins of outbreaks. And yet somehow among the masses of nouveaux experts (who appear to have no knowledge about analytical tools such as the GFT), asking the question about origins constitutes an unacceptable outrage.

SARS-CoV-2

SARS-CoV-2 appears to have started in Wuhan, a large city in Hubei province, China, and a major travel and trade hub. Australian vaccine researcher Dr Nikolai Petrovsky has shown in his research that of all the animal species, SARS-CoV-2 is best adapted to humans. In his vaccine development work, Petrovsky and his team used a supercomputer to analyse all the possible animal relatives of SARS-CoV-2. If the virus came from an animal, then it should have been most well adapted to that species. What they found, however, was that it was by far the most adapted to humans. Not bats. Not pangolins, nor monkeys, nor any other animals. If the virus did indeed emerge from a bat, the researchers should have found it to be more highly adapted to bat species than humans. Petrovsky had enormous trouble getting the research published, with hostile reviews and long delays. Finally, it was published in June 2021.

Professor Shi Zhengli and her team had been hunting for bat viruses related to SARS-1, trying to learn where the virus originated. In 2012, six miners working in the mines in Mojiang, Yunnan, developed severe pneumonia and three died. Shi and her team were sent to investigate. In a 2020 interview in *Scientific American*, Shi attributed the miners' illness to fungus in the mines, but also said that she and her team had identified several coronaviruses in the bats. There is no published scientific evidence that the miners died of fungus, and Shi does not specify what kind of fungus it was. Then in February 2020, Shi Zhengli published a paper in *Nature* showing that a bat virus in her collection from Yunnan, Bat coronavirus RaTG13, had 96 per cent homology to

SARS-CoV-2. Strangely, this was the first time that the virus was reported. A Master's thesis completed at a Chinese university and uncovered by other researchers suggests that the miners' illness was caused by a bat coronavirus, not a fungus. The miners' symptoms and radiological findings were similar to COVID-19, including the classic appearance on chest X-rays of widespread abnormalities that look like 'ground glass'. The Wuhan Institute of Virology has never stated that RaTG13 was identified in bats in the same Mojiang mine, only that it was from Yunnan province. The mine in Yunnan is more than 1500 kilometres away from Wuhan, so how did SARS-CoV-2 evolve from its closest known ancestor, RaTG13, and cause an outbreak thousands of kilometres away – as it happens, very close to the Wuhan Institute of Virology?

It is clear that RaTG13 and other coronaviruses were being studied at the BSL4 lab in Wuhan. We know that the first recognised cluster of cases was in Wuhan, close to the Wuhan Institute of Virology. Shortly before the outbreak, the institute had moved locations, closer to the markets. Is it possible a lab worker became infected from one of the lab's collections of viruses during the move? Could this have been the beginning of the epidemic? When did human infection start? The official narrative still focuses on the wet market and on China. Interestingly, wastewater testing carried out in Spain detected the presence of the virus in March 2019. The finding was mentioned in a preprint in June 2020, but curiously the ensuing publication by the same authors did not report it. Perhaps the test was a false positive or cross-reacted with other coronaviruses. But it is also possible that the journal requested the authors remove all references to it.

As a researcher and a journal editor of four journals, I know that reviewers often request the removal of data that they cannot understand. And in order to get a paper published, the author needs to comply. The wastewater data from March 2019 in Spain is an outlier, outside of the expected bounds of probability. When we analyse any collection of data, we see a distribution of data points, commonly a bell-curve or 'normal distribution', which is a peak with tails on either side in the shape of a bell. Sometimes the shape can be different, but often there is a clustering of data and there may be outliers that sit far from the other data points. As a student of field epidemiology, I was taught that it is common practice to exclude outlier data, because it cannot be easily explained, and it is easier to ignore it. I was also taught that it is important to always study outlier data, because it may be very informative, and tell you about the origin of an outbreak.

So, if the finding from March 2019 was a true positive test, how on earth did it get into the wastewater in Spain a year before the pandemic? In Italy, the virus was documented in a skin sample from November 2019, and its first appearance may extend even further back in time, possibly as early as September 2019. In France, blood samples showed evidence of SARS-CoV-2 infection in November 2019. In the US, we know that serological studies show the virus was circulating in December 2019.

My own research team, using the artificial intelligence–driven system EPIWATCH, which I developed in 2016, showed a signal for COVID in China in mid-November 2019. It also revealed evidence of media censorship, with the term 'SARS' redacted from the headlines of news stories. We found a report of an unusual case of severe pneumonia from Hubei province; the sufferer was flown by helicopter to Wuhan on 17 November. By late 2019, independently, there was another documented supposed 'patient zero' in mid-November, but WHO was only notified on 31 December 2019.

The World Military Games, held in Wuhan in October 2019, provide another interesting line of evidence. If SARS-CoV-2 was circulating in Europe in November, it likely arose prior to that – perhaps in October or even September. The incubation period of the original virus was 12–14 days, so this would be logical. During or after the games multiple competing teams reported an influenza-like illness. The US team flew into Wuhan from Seattle and flew back there. Seattle was Ground Zero for the pandemic in the US. These are intriguing factors. If I myself were in the position of gathering intelligence, I would test stored serum taken between October and December 2019 from any of the athletes who went to Wuhan in October 2019 and see if they have antibodies to SARS-CoV-2. The WHO team did acknowledge the reports of illness at the World Military Games but said there wasn't any evidence that the pandemic was linked to the games. However, they did not present any evidence or test the stored serum from athletes who attended.

Multiple lines of evidence suggest that COVID emerged earlier than December 2019. However, substantial cognitive dissonance, in the form of strong counter-narratives in 2021 and 2022, have tried to shift the focus back to the Hunan wet market and events in December 2019 and January 2020. Instead of rejecting other possibilities, a thorough investigation would consider all lines of evidence.

Let's put ourselves for a moment in the position of someone wanting to deliberately cause a pandemic. For maximum impact, they would pick a major travel hub, a huge city where global transmission would be guaranteed. They would pick a time of year when people are travelling much more than usual – a holiday season perhaps. COVID-19 began in Wuhan, a city of 11 million people which is a major international travel hub. The timing was around the Lunar New Year when everybody was travelling to join their families for the most important holiday in China. The location and timing of the COVID-19 outbreak might have seen it cause extreme damage. Yet the early strain that emerged in Wuhan was controlled well with lockdowns and other public health measures, and the R0 was estimated to be around 2, much lower than later variants. At the very peak of the epidemic in Wuhan, only a few thousand cases were reported per day – far fewer than the 300 000 daily cases in the US or Europe a few months later.

The original strain was less infectious, easier to control and died out by March. The new strain that caused the first wave in Europe and the US in March was D614G, and that was the dominant strain for most of 2020. That variant was more infectious than the original Wuhan strain, but much less infectious than Delta, which has an R0 of 6–8. Early phylogenetic analyses in the US showed multiple introductions of D614G into the country – in New York, Seattle and California, not just a single outbreak starting and then taking off. In 2020 the general understanding was that coronaviruses are much more stable than influenza, and so the expectation was that it would not mutate much. In a complete surprise to most experts, however, the mutation rate has been as high as influenza A, which is highly mutable. Around September 2020, just as vaccines were rolling out, three new variants of concern emerged simultaneously in the UK, Brazil and South Africa. These were rapidly followed by variants in New York, California and Japan, which were more lethal, more transmissible or more vaccine resistant compared to the D614G strain. The new variants were not closely related to D614G but were quite divergent strains. D614G was replaced by the more contagious Alpha, which outcompeted the other variants and dominated until April 2021, when Delta took over, becoming the dominant strain globally for much of 2021. A small number of countries like South Africa and Bangladesh had Beta epidemics instead of Alpha in that time. By November 2021, the Omicron variant emerged, and rapidly became the dominant strain.

When the Omicron variant emerged in November 2021, it stood very far from Delta in the evolutionary tree. In fact, it was more closely related to earlier versions of SARS-CoV-2 than to Delta. When simple Darwinian principles tell us the statistical probability of a new variant being distant from the dominant global virus, Delta, is low, how can we explain the emergence of Omicron? Delta had been spreading everywhere, had outcompeted Alpha and all other variants, everywhere in the world. This meant that statistically it was far more likely that a new variant would emerge from Delta. The same strange phenomenon was also seen when Alpha, Beta and Gamma emerged simultaneously in late 2020, just as vaccines burst onto the scene and promised to save us. All were distant on the evolutionary tree from D614G, which had dominated globally in 2020, and was itself more contagious than the ancestral strain in Wuhan.

We know that gain-of-function research occurs and is even published in scientific journals. In September 2021, Professor Paul Bieniasz and his team from the Rockefeller University, New York, published a paper in *Nature*. As Bieniasz told NPR, 'the goal was to answer the question: Is it possible for SARS-CoV-2 to completely evade neutralizing antibodies?' In order to explore this question, Bieniasz and his team engineered a mutant version of the SARS-CoV-2 spike protein to make it resistant to vaccines. They took about 20 mutations that occurred already around the world, but never all together, and engineered a 'polymutant spike mutant' – in other words, they showed how to make SARS-CoV-2 resistant to vaccines. Two months later, by November 2021, the Omicron variant had emerged, which had most of the 20 mutations Bieniasz and team published with open access methods, plus more.

Meanwhile virologists, acknowledging that on the evolutionary tree Omicron looked an odd outlier, were coming up with theories about where the strain had come from. Some alleged that it had been circulating undetected since the previous year – highly unlikely given that it took four weeks to wipe out the Delta variant globally, how could it have circulated undetected for a whole year? There was also a bizarre theory that it had come from animals – 20 years had separated SARS-1 and SARS-2, but now apparently consecutive animal spill-over events could happen rapidly. That's about as plausible as the frozen food narrative. A more plausible theory held that it had developed in an immunosuppressed person with chronic infection. That is indeed possible, as people with weakened immune systems do not clear the virus as quickly as healthy people do, effectively providing viruses with a live petri dish in which to evolve. But not one virologist postulated that it was engineered, despite it having all the genetic signatures of the super-mutant virus published two months earlier, complete with methods on how to create it. The only thing everyone agreed on was that the origin of Omicron was mysterious.

A smorgasbord of viruses

What about smallpox, the most feared virus? In 2017, for a relatively modest cost, David Evans, a Canadian virologist, and his team synthesised from scratch an extinct virus very closely related to smallpox. The following year, Islamic State of Iraq and Syria (ISIS) issued propaganda warning of biowarfare, using the phrase 'we will make you fear the air you breathe'. The threat seemed to suggest the deployment of bioweapons. This caused alarm in the intelligence community, which had long assumed that smallpox was too difficult to make in a lab. In a 2018 paper, however, Evans and his colleague Ryan Noyce wrote that, 'The advance of technology means that no disease-causing organism can forever be eradicated.' The reality of synthetic biology hit hard after Evans and Noyce's study was published, because it was a recipe for creating an orthopoxvirus, the family of viruses that includes the smallpox and monkeypox viruses. These are DNA viruses, among the largest viruses known, with a much larger genome than RNA viruses. All the previous engineered viruses – polio, influenza, SARS-like viruses – were RNA viruses. Law enforcement and intelligence agencies had been complacent because they had been told creating smallpox synthetically was too technically difficult, but Noyce and Evans blew that theory out of the water.

PLOS Pathogens, the journal that published Evans and Noyce's paper, also published an editorial defending the paper. In it they pointed out that the scientists used known methods that had already been utilised for creating polio, influenza and even vaccinia virus, to conduct the research. The defence was essentially a straw man argument – after all, almost all scientific research builds on past research and the authors themselves acknowledged the dual-use risk. The Evans and Noyce experiment was published in an open access journal, meaning the methods are available forever to anyone who cares to use them.

Soon afterwards, in 2019, the US CDC synthesised Ebola from scratch and published the methods. Later, Evans published a paper showing the exponential decline in the price of synthetic biology research. Synthetic biology is easily accessible and cheap – the cost has decreased 250-fold in a decade, so that cost is no longer a barrier to nefarious use of the technology. There are over 200 unregulated private companies in synthetic biology, governed only by a voluntary code of conduct to report nefarious activity. In 2009 the commercial gene synthesis companies formed the International Gene Synthesis Consortium (IGSC). The consortium functions 'to design and apply a common protocol to screen both the sequences of synthetic gene orders and the customers who place them. In addition, the consortium works with national and international government organizations and other interested parties to promote the beneficial application of gene synthesis technology while safeguarding biosecurity.' Nonetheless, this is still a self-regulated industry with vested interests at play, and the IGSC does not represent all the players in the field. Nor does it make its database of regulated pathogens open source or public.

The potential consequences of all of this are alarming. If terrorist groups, nation states or any other group wanted to use Ebola or smallpox, they could order sequences for dangerous pathogens and feasibly make them themselves for use as a weapon.

6

THE SELF-REPLICATING WEAPON

BIOLOGICAL WARFARE HAS BEEN USED THROUGHOUT recorded history. It dates back at least to 300 BC, when the Greeks, Romans and Persians used cadavers to contaminate the water supplies of their enemies. Centuries later, in North America, smallpox was introduced to Native Americans in contaminated blankets by Lord Jeffery Amherst to decimate their population during the Pontiac's War in 1763. There is plenty of documented evidence demonstrating that this was intentional, including written notes from Amherst in which he wrote:

> Could it not be contrived to send the small pox [*sic*] among the disaffected tribes of Indians? We must on this occasion use every stratagem in our power to reduce them ... You will do well to try to inoculate the Indians by means of Blankets, as well as to try Every other method that can serve to Extirpate this Execrable Race. I should be very glad your Scheme for Hunting them Down by Dogs could take Effect, but England is at too great a Distance to think of that at present.

It is likely this method was also used against Indigenous Australian populations. In 1789, a year after the arrival of the First Fleet, an epidemic of smallpox devastated Aboriginal resistance in Sydney and surrounding areas, killing over 75 per cent of the local population. Some have argued that the outbreak was, in fact, chickenpox. But Australian scientist Dr Frank Fenner, a world-leading smallpox expert, was convinced it was smallpox.

Born in Ballarat in 1914, Fenner was an army doctor and virologist, with a distinguished career at both the Walter and Eliza Hall Institute in Melbourne and the Australian National University in Canberra. Chairman of the Global Commission for the Certification of Smallpox Eradication and a leading light in the WHO eradication campaign, Fenner argued that epidemiologic evidence of immunity in people with the telltale scars of smallpox during a later epidemic in 1829 was proof that this was smallpox. Australian Aboriginal peoples had been free of measles, smallpox and many other vaccine-preventable infections until the First Fleet arrived, so the only source of immunity during the epidemic 40 years later would have been from the initial 1789 epidemic. Further, the journal of a captain on the First Fleet documented that bottles containing smallpox were carried on the ships. A range of counter-narratives have been thrown around, but the most credible explanation is a deliberate attack at a time when supplies were dwindling, and the relatively small forces of the occupiers had difficulty dealing with both the transported convicts and the resisting Aboriginal population. The epidemic occurred at a distance away from the First Fleet settlement in a position consistent with release from boats on the harbour. Researcher

Chris Warren, who has studied this extensively, concludes that the history of Australia may have been very different had smallpox not been used.

Another instance of biological warfare is the Second Sino-Japanese War (1937–45). Like the First Fleet and First Nations people, the Japanese were vastly outnumbered by the Chinese. The Japanese used biological weapons as a strategy to gain an advantage. Famed for their biological and chemical warfare expertise, they released plague bombs in Manchuria, producing devastating mass illness and death. They also conducted attacks with anthrax, and tested plague, cholera, smallpox and botulism on prisoners of war. Other techniques included poisoning of water supplies and dropping contaminated food and clothing into free parts of China. It is estimated over 10 000 Chinese were killed by Japanese bioweapons.

Unit 731 of the Imperial Japanese Army was notorious for its bioweapons program, so much so that following the Second World War, the Soviets captured the unit and kept high-ranking officials in a special, lenient prison camp to learn what they could for the purposes of developing their own bio-weapons program. The Soviets themselves had a long history of offensive biological weapons development, starting with the Leningrad Military Academy in 1928. They used tularaemia against German troops in 1942. Thousands of German and Soviet troops in the Crimea died of the disease, with a 70 per cent incidence of the pneumonic form, indicative of a deliberate attack. The Soviets' first major bioweapons program was established at Sverdlovsk in 1946, followed by the formation of Biopreparat in 1973, where they experimented with Marburg, smallpox, anthrax, Ebola, Lassa fever and plague. After the fall of the Soviet Union in 1991, the US began the Cooperative Threat Reduction Program. Initially Russia was involved in the program, but it pulled out in 2012. Today, the US continues to support biolabs in Kazakhstan, Georgia and Ukraine. The US government has links with at least 26 labs in Ukraine and has provided direct support to six of these.

When Russia invaded Ukraine in February 2022, Russian news outlets insisted that their troops were trying to dismantle nefarious US biolabs doing clandestine biological weapons research that they would not be allowed to do in the US. On the other hand, Western press agencies have reported that the Russian invasion poses a risk for the accidental release of dangerous pathogens stored in these labs. Some pathogens are stored in frozen form, meaning that a loss of power through the disruptions of war could convert any inert, frozen pathogens into viable ones. In addition to causing massive damage and significant casualties, the Russian invasion of Ukraine has resulted in close calls around nuclear power plants. The invasion could just as easily have resulted in attacks on laboratories. In fact, some social media have stated that such attacks have already taken place.

One news site has claimed that 'Putin has ordered his military to seek and destroy US-Deep State biolabs engaged in top-secret zoonotic and infectious disease research in dozens of locations across Ukraine.' Yet according to Andrew Weber, a board member of the Arms Control Association and former US Assistant Secretary of Defense for Nuclear, Chemical and Biological Defense programs, the Russians have used a disinformation campaign to justify the invasion, and that the

US role has only been to provide 'technical support to the Ukrainian Ministry of Health since 2005 to improve public health laboratories whose mission is analogous to the US Centers for Disease Control and Prevention'.

Even if the latter is true, the US government's Defense Threat Reduction Agency (DTRA) can get around accountability to the US Congress by outsourcing lab work to private companies who can operate with greater freedom. The Richard M. Lugar Center for Public Health Research in the former Soviet state of Georgia, has a virology lab with a global health security remit and an official US Army Medical Research Directorate. Private defence contractors including CH2M Hill, Battelle and Metabiota, have been awarded contracts from DTRA to do work there. Metabiota also has contracts in Ukraine. The US Embassy in Ukraine has stated that the US Biological Threat Reduction Program's 'priorities in Ukraine are to consolidate and secure pathogens and toxins of security concern and to continue to ensure Ukraine can detect and report outbreaks caused by dangerous pathogens before they pose security or stability threats'. This means dangerous pathogens are contained within and could be accidentally released during conflict whether or not bioweapons research is occurring in these labs. It should be no surprise that many countries are conducting bioweapons research. Some of this is defensive research to prepare for an attack. But some may also be offensive research.

Offensive bioweapons programs

Documented offensive bioweapons programs have existed in many countries, including the United States, the USSR, the United Kingdom and South Africa. During the apartheid era, although being a signatory to the Biological Weapons Convention, South Africa engaged in a major offensive chemical and biological weapons program. Known as Project Coast, the program was headed by a medical doctor, Wouter Basson, who was private physician to President P.W. Botha. South Africa had two main bioweapons facilities: the Institute of Virology in Johannesburg and the South African Defence Force veterinary lab near Pretoria. These labs developed weapons specifically to kill Black Africans, including nerve agents, cyanide, anthrax, brucella and cholera. They even used chemical agents to sterilise men and conducted research on reducing the birth rate in the Black population using clandestine contraceptives in the water supplies. In 1981 Basson visited the US to meet with American biological warfare experts. Colonel Johan Theron testified that he killed over 200 prisoners with chemical agents provided by Dr Basson and dumped their bodies in the sea. Basson was alleged to have been involved in many other medical killings and was later even questioned about illicit trade in weapons and nuclear material. But in 2002 he was acquitted of all charges. Then in 2006, the Health Professions Council of South Africa investigated Basson and found him guilty of unprofessional conduct. In 2019, the High Court overturned this decision. Today, Basson continues to practise medicine and tour the world as a celebrity speaker.

Like South Africa, Rhodesia too had a chemical and biological weapons program. The program was used against the Zimbabwean African National Liberation Army (ZANLA), probably from 1972 onward, escalating after the Portuguese withdrew from Mozambique in 1976. In the 1970s Rhodesia turned to 'pseudo operations', creating the Selous Scouts in 1973. South Africa is believed to have used Rhodesia as a beta-testing ground for its own bioweapons program. Rhodesia's Selous Scouts, which operated under the Central Intelligence Organisation (CIO), used injured or captured Black Zimbab-weans to serve as spies and provide intelligence. South Africa is thought to have played an important role in assisting Rhodesia, including funding the Selous Scouts in exchange for access to the military base at Mount Darwin. Rhodesia also developed a bioweapons program under Professor Robert Symington. The biowarfare included poisoning of water and food, spreading cholera, infecting clothing and using anthrax to kill cattle. They stockpiled ricin, thallium, organophosphates such as parathion, warfarin and anthrax.

In 1975 Rhodesia began clinical trials of biological and chemical agents on prisoners in a Selous Scouts secret detention centre in Mount Darwin. Selous Scouts posing as insurgents injected thallium into tinned meat and distributed it to real insurgents. They also injected cyanide and other poisons into bottled alcohol. Poisoning of water supplies was common. In one instance, over 200 locals died when a well, which was their only water source, was poisoned. In another instance, a cholera epidemic began in Mozambique when the Ruya River, which flows between Zimbabwe and Mozambique, was deliberately contaminated. The CIO also posed as guerrilla recruiters and provided recruits with clothing and uniforms poisoned with organophosphates. In 1978 they killed large numbers of rebels with the blood thinner warfarin – at first the bleeding illness was thought to be a viral haemorrhagic fever but was then diagnosed as warfarin poisoning. Throughout these experiments, dead bodies were thrown down mine shafts to get rid of the evidence. In 2011 mass graves with hundreds of bodies were found in these mines.

In 1978, at the peak of the Rhodesian struggle to maintain power and just after their invasion of Mozambique and Zambia, the largest known epidemic of anthrax occurred in Tribal Trust Lands in Rhodesia. The effects of the outbreak on human and animal life were devastating. Official reports stated that 10 783 people became infected and at least 200 died, and huge numbers of cattle perished. A painstaking analysis of the data of individual reports by researchers in the US showed at least 17 199 human cases, and that official reports were an underestimate. The outbreak, which continued until 1980, was largely restricted to the Tribal Trust Lands – arid, less productive areas owned by Black farmers. White farmers, who owned the best farming land, were mostly spared. Prior to 1978, natural anthrax, endemic in many parts of Africa, had been rare in Rhodesia, with only a handful of cases each year. Between 1950 and 1978 there had been a total of 334 cases. Analysis by American bioweapons expert Dr Meryl Nass concluded that the 1978 anthrax outbreak in Rhodesia was likely biowarfare. Nass based her assessment on the geographic clustering, unprecedented scale and the fact that most of the Tribal Trust Lands across six of eight provinces were affected. Further, many cases occurred in areas where there had never been any anthrax previously, and neighbouring counties were unaffected.

The timing of the anthrax outbreak with the peak of the Rhodesian Bush War is also suggestive. It is likely that the Selous Scouts were involved, along with Rhodesian psychological operations and the CIO. They would have had three core objectives: to kill any person who might otherwise have fought with the ZANLA, to kill cattle in order to destroy food supply and cause starvation, and to cause economic harm. In rural areas, wealth was measured by the number of cattle owned by people. In an interesting twist, Dr Steven Hatfill, the first suspect investigated by the FBI in the 2001 anthrax attacks in the United States, had lived in Rhodesia from 1978 to 1994 and claims to have joined the Selous Scouts. This and another piece of circumstantial evidence made the FBI initially focus on Hatfill. In the end, however, no firm evidence was found against Hatfill, and the FBI moved on to Bruce Ivins.

Bioterrorism

When a nation state uses biological toxins or infectious agents with the intent to kill, harm or incapacitate humans, animals or plants, this is called biowarfare. If the same act is perpetrated by a non-state actor, it is called bioterrorism. Numerous cults and terrorist groups have engaged in biological attacks. These have included the American cult Order of the Rising Sun, German militant group Baader Meinhof, and the Rajneesh cult. In Japan in the 1990s the Aum Shinrikyo cult attempted aerosol release of botulinum toxin three times without success in the Tokyo subway. In 2014, ISIS publicly executed the head of the Physics department at the University of Mosul in northern Iraq for refusing to develop bioweapons for them. A number of lone wolf actors have also attempted biological attacks. These include white supremacist Larry Wayne Harris, a trained microbiological technician from Lancaster, Ohio, who in 1995 ordered plague by mail order. After a suspended sentence for this, he was arrested again in 1997 for a plot to release plague in the New York subway. In 2014 a New York University student, Cheng Le, attempted to buy ricin on the dark web. In 2020 US Coast Guard and white supremacist Christopher Hasson was jailed for stockpiling weapons and researching biowarfare for the purposes of an attack.

Whether it's called biowarfare or bioterrorism, the end result is the same. The US CDC defines bioterrorism as 'the deliberate release of viruses, bacteria, or other germs (agents) used to cause illness or death in people, animals, or plants'. Interestingly, there is no universally agreed definition of terrorism. The term was first used in France to describe the systematic use of terror by the French government against its own people in 1793. Since then, the term has changed its meaning, and today is often used to describe actions of non-state actors against states. It remains subjective depending on perpetrator and victim and has both political and legal meanings. Most definitions include the use or threat of serious violence to elicit public 'terror' (as opposed to domestic violence or private terror). However, most historical biological attacks have been carried out by stealth and creating terror has not been a motive. For example, the Rajneesh case was a covert attack with the motive of controlling a local election.

In 2018, ISIS released propaganda threatening biological attacks on their enemies, and issued posters with the tag line 'We will make you fear the air you breathe.' Knowing that at least one terrorist group has a motive and stated intent poses a real threat. The means are readily available for clandestine biolabs to be set up, with easy access to DIY biology materials and methods, and there is no feasible way of tracking them or ensuring effective oversight of them. In the Rajneesh case from 1984, it was only by chance that the FBI discovered the biowarfare lab, which contained many other, more dangerous pathogens. Clandestine labs now have much greater capability than the Rajneesh lab had nearly 40 years ago. At the same time there is also a greater risk of accidents or leaks.

Most bioterror agents such as anthrax, plague, tularaemia or Ebola also occur in nature. This may make it difficult to determine natural or unnatural origin of an outbreak. A range of more common infections, including influenza or salmonella, can also be used when stealth is the objective. Most commonly, the bioweapons of choice are bacteria or viruses. Bacteria, which are larger than viruses, can cause disease by direct invasion, toxin production or through spores (as in the case of anthrax). Examples include staphylococcus, streptococcus, anthrax, E-coli, meningococcus, salmonella, tuberculosis and plague. Most bacteria can be treated by antibiotics, and some can be prevented by vaccines. Some bacteria, however, are resistant to most known antibiotics. These may have a high death rate and include Methicillin-resistant Staphylococcus aureus and XDRTB (extensively drug-resistant TB). If bacteria are used as bioweapons, drug resistance can be engineered.

Viruses, in contrast to bacteria, are much smaller, and are intracellular. Viruses have a protein coat and core of genetic material which is either DNA or RNA. Unlike bacteria, which can replicate in many different environments such as food or soil, viruses require living cells in which to replicate. They cause disease by invading human cells and then using our cells to reproduce themselves. Some antivirals are available (for example, for influenza, HIV and now Paxlovid and Molnupiravir for COVID-19), but most viruses have no treatment. We do, however, have vaccines for many viruses, including measles, mumps, rubella, polio, hepatitis A and B, chickenpox, influenza, COVID-19, Ebola and smallpox. It is possible to engineer vaccine resistance for viruses, as in the case of the accidental resistance to smallpox vaccines which occurred in the Australian mousepox experiment.

Signed in 1972 and implemented in 1975, the Biological Weapons Convention (BWC) was the first multilateral disarmament treaty banning an entire category of weapons. A supplement to the 1925 Geneva Protocol prohibiting the use of chemical and biological weapons in war, the convention was signed by 22 countries, including the US and the USSR. The BWC, however, is neither enforceable nor auditable, despite undergoing multiple reviews and revisions since 1975. It was born of the Cold War, and, as such, it assumes the major threat is from nation states. This is no longer the case, given the wide array of possible bad actors and motives, as well as the ease of access to biological agents. The BWC is not a fit framework for contemporary terrorism threats, because its focus is on nation states. Biological attacks can and have featured perpetrators who were neither a nation state nor a terrorist group – the

2001 anthrax 'letter-bombs' attack in the US stands out as a perfect example. Since the September 11 attacks, the world has been transformed in unanticipated ways, with terrorism now part of the global landscape. The so-called 'war on terror', launched by the US government in the wake of 9/11, has now been running for a longer period than the First and Second World Wars combined, with no end in sight.

A biological bomb

Like guns, biological weapons can be used against individuals. They can also be directed against whole populations, like bombs. Unlike bombs or guns, however, the harm caused by biological weapons is not immediately apparent. After a biological weapon is deployed, depending on the incubation period of the infection, it may take days or even weeks before the disease manifests. The weapon itself, a microorganism, is invisible. In contrast to a bomb or gun, a bioweapon used against one person could infect others who were not targeted by the attack. This makes bioweapons unique. Whether an epidemic arises after a bioweapon attack depends on the mode of transmission and how contagious it is. Some bioweapons are not transmissible from person to person, so the resulting outbreak does not spread any further. This pattern, called a 'point source' outbreak, is typified by the 1984 Rajneesh salmonella attack or the 2001 US anthrax attack.

Bioweapons that spread from person to person have the greatest potential as weapons of mass destruction – multiplying and spreading from host to host, they are the gift that keeps on giving. The particular way that a virus spreads from person to person matters too. For example, a virus that is sexually transmitted (such as HIV) does not have the same potential as a weapon of mass destruction as a virus that spreads through the air we breathe. The most feared weapons are those that either have a very high fatality rate (like anthrax or Ebola) or are spread through the air from person to person and have pandemic potential. However, in terms of ease of spread, microorganisms transmitted by contact (touching) are the easiest to deploy – the perpetrator simply has to contaminate a surface that will be touched, like a door handle. Or if mass infection is the aim, high-touch surfaces at a large public venue such as an airport, sporting event or train station, can serve the purpose.

In terms of locations for mass attacks, in order to achieve maximal impact, the most likely are places where large numbers of people congregate – airports, train or bus stations, city centres, schools, shopping malls and other crowded settings. Airports may be favoured, as dispersion of travellers around the world would ensure maximum spread and impact. Another factor to consider is the dose-response relationship. This is a known phenomenon where, the higher the dose of exposure, the shorter the incubation period, and the more severe the symptoms and illness. A deliberate attack would likely involve much higher doses than natural exposure, and therefore may be characterised by more severe illness and quicker onset. For any infection, there is an expected distribution of severity from asymptomatic at one end to death at the other. Clusters of infection that show a higher-than-expected rate of severe infection and a shorter mean incubation than expected could suggest a potential bioweapon is at work.

The most feared weapons are those that are airborne. These require aerosolisation – the process of making the virus spread through tiny, microscopic particles, often carried on water, that are light enough to float in the air – which is more technically complex than contamination of surfaces. Some viruses are naturally airborne, like smallpox or SARS-CoV-2, which are breathed out in tiny, microscopic particles of fluid that come from the lungs during normal breathing. Others, like anthrax spores, can be coated with silicone or other substances to enhance airborne potential. During the Cold War, crop duster planes were considered a likely method for an airborne attack. Such methods would be easily identified by radar and other techniques to detect planes or maritime vessels. Today, other approaches would be more likely to be used. Aerosolisation can be more easily achieved through air-conditioners, ventilation systems or cooling towers than through a customised weapon or an aircraft. The success of airborne transmission depends on several factors, including how hardy the pathogen is in the air and the environment. Some pathogens are very robust, and can remain viable for days, weeks or even years. Others are destroyed within minutes in the air. If a virus only remains viable for 30 minutes in the air, this allows only a short window for the timing of attack. Influenza can persist up to three hours in the air, and SARS-CoV-2 up to 16 hours. Smallpox too can persist for hours in the air, and can remain preserved in scabs for many years.

Asleep at the wheel

As far back as 2006, I published papers on critical infra-structure dependencies, surge capacity and all the domino effects of mass infection during a pandemic. I showed health systems' capacity and infrastructure planning are as equally important as drugs and vaccines. My research published in 2019 looked at the impact of a bioterrorism attack with smallpox on the health system in Sydney. Using a mathematical model, we determined requirements for hospital beds, contact tracing and healthcare workers in Sydney – long before most people had even heard of contact tracing. We showed that the requirement for extra hospital beds could range from 4 per cent to 100 per cent of all available beds in best- and worst-case scenarios. I warned that in a severely stressed health system, care of other urgent patients, such as people with heart attacks or strokes, may be affected due to a shortage of beds and staff. The surge in beds requires a corresponding surge in doctors and nurses to treat infected patients, as well as adequate PPE for these clinicians. We pointed out that most stockpiling provides short-term supplies, but a pandemic can go on for months or years. In this case, standard stockpiles (which typically provide a few weeks' supply) will be rapidly depleted.

Vaccination is critical for epidemic control, but the process of vaccination requires more than the vaccine alone. Smallpox control using ring vaccination (finding and vaccinating close contacts of each case of the disease) requires contact tracing, with the number of contacts expected to be an order of magnitude greater than cases. Yet the available workforce in public health to conduct contact tracing is much smaller than the clinical workforce. In our research study, we estimated that while there are about 100 000 clinical healthcare workers in Sydney, there are only about 300 public health workers. Public health workers are trained in disease surveillance, contact tracing, epidemic control, health promotion and other aspects of population health. They usually have done a Master of Public Health and work in health departments or public health units. To make matters worse, trained public health workers are not registered as health practitioners in Australia. This makes it difficult to identify the available pool of qualified workers. In our modelled smallpox epidemic, we said it was likely that community volunteers would be required to assist with contact tracing, as was the case during eradication of smallpox. And that has been exactly what has happened with COVID-19. In 2020, airline staff were recruited to carry out contact tracing phone calls to meet the surge in demand for contact tracers. We warned that countries like Australia without civil preparedness programs or national service would be more vulnerable if community volunteers are needed for surge capacity. I believe that there is a strong case for offering civil preparedness courses, training and certification to enable rapid community surge capacity.

Testing and identifying new cases is vital because isolation stops infected people infecting others. Contact tracing is important because the close contacts are the next most likely to become infected. If they are identified early and quarantined, they won't infect others. If testing, tracing, isolation and quarantine (TTQI) protocols are not observed, exponential growth occurs, because each person has 10–20 contacts on average. For a virus with a reproductive number (the number of infections arising from one infected person) of 6–8 like the Delta or Omicron variants of SARS-CoV-2, this means that one person infects at least six others, who then go on to infect 36 more, who, in turn, infect 216 more. In such circumstances, an epidemic grows extremely rapidly. Vaccine, masks and TTQI will all dampen that growth and flatten the curve. If nothing is done or mitigation is minimal, or, worse, if infectious people are allowed to go to work (as we are seeing around the world today), rapid epidemic growth will overwhelm hospitals and the health system.

Designated hospitals

In the case of a large-scale epidemic of a serious infection like Ebola or smallpox, it is important to have plans for designated or pop-up treatment centres (rather than treating cases in any general hospital). This not only reduces the risk of hospital outbreaks but can reduce the number of staff exposed to infection and enable better patient cohorting (separating of infected and uninfected) and rostering of staff. During the Ebola epidemic in 2014, the Royal Australian College of General Practitioners (RACGP) released a poster telling people that if they thought they had been infected with Ebola, they should go to their GP. Together with an academic GP colleague, I wrote to the RACGP and advised them this was a very bad idea – the last thing you want during an outbreak of a serious infection like Ebola is for people to present willy-nilly to GPs all over the country. This will disseminate the infection rapidly. It's a basic principle that has been followed for centuries: minimise the number of sites handling the infection by having dedicated hospitals for the condition. This was adopted, for instance, with smallpox hospitals in the UK in the nineteenth century, including smallpox hospital ships on the River Thames. Fortunately, in the end, the RACGP withdrew the posters, and a small number of hospitals were designated for Ebola.

In the last European smallpox epidemic in 1972, hotels and motels were commandeered on the outskirts of the city to use as treatment and quarantine facilities. In 2014, after Ebola caused an outbreak in the First Consultants Hospital, Nigeria did well to control the epidemic. The response team drew on extensive experience with polio control campaigns, as Nigeria had been one of the last strongholds of polio in the world. They immediately established a separate designated Ebola treatment centre in an abandoned building and moved patients out of the general hospital environment. Aside from the initial missed diagnosis, Nigeria did an exceptional job of containing and stopping the hospital epidemic and preventing a larger scale epidemic in the community.

A large-scale epidemic in the capital city of the most populous country in Africa would have been a disaster. One surviving doctor, Ada Igonoh, wrote an account of the management of the epidemic in Nigeria through her personal experience of becoming infected and being taken to a makeshift facility for Ebola care. It was an abandoned building, and no one came in to care for them, except to drop off food and check in on them. She knew she had to keep hydrated and have electrolytes to replace the loss of fluid and electrolytes from the diarrhoea and haemorrhaging, so forced herself to drink the oral rehydration solution provided. Several of her infected colleagues died around her in that place, but she survived. When she left, she described the meticulous infection control measures taken, including on exit from the facility. She had to leave everything behind including her clothes, iPad and phone, and take a chlorine bath before changing her clothes. These processes no doubt prevented further spread of Ebola in Nigeria.

A bioterrorist incident with a serious virus like Ebola or smallpox will likely see us stumble our way through, missing the initial diagnosis, allowing spread, and learning as we go. As we've seen, past experience demonstrates how easily this can happen: the diagnosis of smallpox was missed in Yugoslavia in 1972, and Ebola was similarly missed in Nigeria in 2014. In 2022, the monkeypox outbreak is thought to have been quietly spreading undiagnosed for some time. Its re-emergence is linked to waning immunity from smallpox vaccines (which can also offer protection against monkeypox), last used in whole populations in the 1970s before the eradication of smallpox was declared in 1980. Australia never used mass vaccination, and we estimated in our research that only 10 per cent of Australians are vaccinated. We also live with far greater levels of immunosuppression today due to advances in medicine for the treatment of cancer and autoimmune diseases. With people under the age of 50 likely never to have been vaccinated, and with immunity waning in older vaccinated people, smallpox may be more severe if it emerged today.

7

THE SPECTRE OF SMALLPOX

LIKE SMALLPOX, MONKEYPOX FORMS PART OF THE orthopoxvirus genus of viruses. These are closely related viruses and include the vaccinia virus that is used as the vaccine to protect against smallpox. Monkeypox is the most serious current orthopoxvirus threat. In 2022 an inexplicable outbreak exploded in the UK and Europe, spreading to over 70 countries around the world prompting the Director-General of WHO, Tedros Adhanom Ghebreyesus, to declare a Public Health Emergency of International Concern in July 2022. The Director-General acted because the emergency committee could not reach consensus on declaring an emergency. They met first in May and again in July, and both times failed to reach consensus.

Prior to 2017, monkeypox was a rare infection, endemic in seven African countries and usually spread from animals to humans with occasional human to human spread. It has been causing very large outbreaks in Nigeria and the Democratic Republic of Congo since 2017. Scientists have puzzled over why a previously rare infection is now becoming more common.

I began researching this question in 2017. One possible cause is the waning of immunity from smallpox vaccination, confirmed by our monkeypox research. We also showed that blood samples from Australian blood donors had almost no immunity to smallpox. In September 2018, a case of monkeypox occurred at a naval base in Cornwall in the UK in a person who had travelled from Nigeria. Simultaneously, a second case occurred in Blackpool on the northwest coast of England in an unrelated person also returning from Nigeria. I was alarmed, because in 2017, there had been fewer than 200 cases of monkeypox in Nigeria, and in 2018, only 45 confirmed and 114 suspected cases. The statistical probability of such a small number of cases in the source country resulting in two unrelated imported cases in the UK at the same time, is extremely low. Was there much more monkeypox that wasn't detected? A third case occurred in a nurse who had treated the Blackpool case, with no further cases until 2022. Two further cases of monkeypox were imported through travel to Israel in 2018 and Singapore in 2019, with no warning of what was to come.

The size and pattern of the 2022 epidemic and rapid spread left experts scratching their heads. The only way they could explain the strange occurrence of multiple unrelated clusters in the UK was by vast numbers of asymptomatic or mild, undiagnosed infection also occurring. We know that smallpox does not transmit in asymptomatic people, so why would monkeypox be so different? Occasionally, people who have had smallpox vaccine may get asymptomatic monkeypox, but not at a scale that could explain the epidemic. It is also possible that large numbers of cases have been misdiagnosed as other infections, such as herpes simplex, but this is not likely to explain the current epidemic. Clusters have occurred mostly among men who have sex with men – strange for a respiratory virus that has not had this pattern before. The spread is consistent with sexual or close contact transmission. There is also an unusual clinical picture with genital rash rather than the classic monkeypox rash on the hands and face. There are over 50 mutations in the virus, the significance of which is yet unknown. Phylogenetic analysis suggests rapid and continuous mutation, which is odd for a stable DNA virus. Key questions are whether the epidemic will spread to the general community and to children – the risk of dying from monkeypox is much higher for children. In countries that have never experienced monkeypox before, it may also become endemic in local animals and therefore a permanent risk.

When the UK epidemic began, alarmed, I let various people in the Australian Department of Health know I was a member of the WHO SAGE smallpox and monkeypox advisory group and stood by, ready to assist with Australia's response. I had published on monkeypox prior to 2022, and sent them a list of my smallpox and monkeypox publications over my 15 years of researching orthopoxviruses. There was radio silence from the department. Later, as the epidemic escalated, I heard the chair of a key committee say at an open forum on vaccines that they knew nothing about orthopoxviruses and had to read up on smallpox vaccines.

A smallpox bomb

Monkeypox is not the only orthopoxvirus of concern. In 2015, a brand new orthopoxvirus, Alaskapox, was identified (so-called because it appeared in Alaska). It presented in two people, and between 2020 and 2021 four cases were identified. Alaskapox appears to spread from animals to humans but is not contagious between humans. It is another virus that we are watching closely. Smallpox, however, remains the most feared orthopoxvirus.

In his book *Biohazard*, former deputy director of the Soviet bioweapons program Ken Alibek says as far back as the 1970s the Soviets were determined to make smallpox even more deadly. They inserted genes from Lassa virus into the smallpox virus, variola. They tested dangerous chimeric viruses on monkeys and rabbits on a little island in the Aral Sea, Vozrozhdeniya, which is now an uninhabitable, toxic wasteland split between Uzbekistan and Kazakhstan. They called the island testing facility Aralsk-7. In 1972, the Soviets exploded a smallpox 'bomb' over Vozrozhdeniya. A ship passing by 15 kilometres away, the *Lev Berg*, sailed through a cloud of the aerosolised smallpox. One of the ship's crew, a technician, got smallpox. But when questions were asked, the Soviet government covered up the incident, heaping the blame on the victim. Official Soviet reports claimed that the infected crew member had got off the ship at several stops, although she herself denied it and the formal ships' policy precluded it. She had been vaccinated, but developed severe illness, suggesting a high dose of exposure. After being transferred to the small nearby city of Aralsk, she infected several others, who manifested a much higher than expected rate of haemorrhagic smallpox, the worst type and universally fatal. This suggests the smallpox bomb utilised a modified, weaponised strain. American scientists referred to this as the Aralsk strain, but their counterparts at the Vector Institute in Russia have always denied the existence of the strain and refused to produce samples.

It wasn't until after the collapse of the Soviet Union that the narrative on what had taken place began to change. The former Deputy Health Minister, General Dr Pyotr Burgasov, revealed that a smallpox 'bomb' had indeed been exploded on Vozrozhdeniya at that time, likely expelling a plume of smallpox that reached the *Lev Berg* as it sailed at sea several kilometres away. While there are many other examples of long-distance transmission of smallpox, at distances of a kilometre or more, none had ever spread as far as this one to the *Lev Berg*. This case thus demonstrated the unique ability of smallpox to spread by an airborne route over a great distance. In the era of smallpox, the virus was accepted as highly airborne.

As a disease, smallpox has been ravaging the earth for centuries. It is caused by an ancient virus, variola, found in Egyptian mummies from the 3rd century BC. Over the centuries, smallpox has caused recurrent epidemics. Like COVID-19 and measles, which are also epidemic infections, smallpox displayed the typical repeating cycle of waxing and waning incidence. Once infected, unvaccinated people had a 30 per cent chance of dying of smallpox – much higher than the 12 per cent mortality rate for SARS and 2 per cent for SARS-CoV-2. In the last 100 years of its existence in the world, before it was eradicated in 1980, smallpox killed over

500 million people. After the introduction of a vaccine, it took the WHO 30 years to eradicate the disease. The Pan American Health Organization started the ball rolling, managing to eliminate smallpox from most of the Americas by the 1950s. This was followed by the WHO eradication campaign which began in 1967 and was completed 13 years later. The last naturally occurring case of smallpox was in 1977, but in 1978 a leak from a UK lab resulted in the infection and subsequent death of Janet Parker.

It was in 1881, after vaccination had just about eliminated smallpox from the community, that the British first noticed the ability of variola to spread over long distances in the air. They called it 'aerial convection'. Previously, smallpox had been so widespread in the community that it was impossible to pinpoint the source of infection when one individual became infected. Smallpox was quite simply everywhere – just like COVID is now. In the last 100 years of smallpox, however, with the incidence much lower after the success of vaccination, unusual transmission events became very apparent and therefore easier to identify. Following the introduction of the smallpox vaccine, the British noticed a recurrent phenomenon of infections occurring in the community within a one-mile radius of the smallpox hospitals, in the absence of any community transmission. Such outbreaks were observed in London, Glasgow, and even around the smallpox ships moored in the River Thames which offered an alternative to hospitals, to try and separate people with the disease from the community. Despite their isolation, these floating smallpox hospitals still resulted in spread to surrounding communities, with the highest number of cases in towns close by, and some cases even appearing in towns a little further away – all without any direct contact. Similar observations were made in Germany, in the United States, Greece and Sweden, at distances of between half a mile or more. The British even brought in a decree that smallpox hospitals must be spatially separated from other buildings and from communities to protect people from aerial convection.

When I began working on smallpox research in 2006, I came to realise that one of my most beloved mentors from the early 1990s, Dr Mike Lane, was a smallpox expert. I learnt more from him about epidemic response than anyone, but only realised his credentials in smallpox much later. Unbeknownst to me, he had been director of the Global Smallpox Eradication program at the US Centers for Disease Control (CDC). Mike had come from the US CDC in 1991 to establish the Australian Field Epidemiology Training Program (also known as the Master of Applied Epidemiology, or MAE) at the Australian National University, and I was in the second cohort, starting in 1992. I had seen an ad for the program while doing my physician training at Royal Prince Alfred Hospital (RPAH) in Sydney and applied because I wanted to learn epidemiology. Professor Bob Douglas, who established the program at ANU, interviewed me at the RPAH cafeteria and I took a two-year, life-changing detour to do the MAE. I was placed at the Health Department in Victoria, in the Collins Street building, where the head of communicable diseases smoked in his office (workplace smoking laws would come a few years later). When Mike left, Professor Aileen Plant, a legend in pandemic control, took over, and went on to be my PhD supervisor, close friend and mentor.

In 1972, Mike had been sent from the US CDC to assist with the last European epidemic of smallpox in what was then Yugoslavia. By 1972 smallpox had not been seen for more than 30 years in Yugoslavia, so the diagnosis was missed. When a return traveller from the Middle East presented to hospital ill, they did not diagnose him correctly. He had haemorrhagic smallpox, the very worst kind with close to 100 per cent fatality. This is where the smallpox rash becomes confluent and haemorrhagic. Thinking that the patient had a severe reaction to the antibiotics, the doctors called in many medical students and other juniors to have a look at the 'interesting rash'. Even the autopsy was attended by many students. A large epidemic resulted, and the diagnosis was only made when the next wave of infections began about two weeks later and subsequent patients had the more classic smallpox rash.

I showed in my research that rapid diagnosis of smallpox is critical. However, delay of diagnosis is common, and many outbreaks of smallpox have not been recognised until second generation cases occurred. In 2017, I interviewed Mike for a podcast and asked him for details about the 1972 outbreak. I am so glad I did, because when he passed away in 2020, some parts of his lived experience and knowledge were captured on that podcast. He told me that in Yugoslavia, after realising it was smallpox, the health authorities did a tremendous job in managing the epidemic with the pillars of epidemic control: test, trace, quarantine and isolate – or TTQI, as we have come to call it during the COVID-19 pandemic. In our research, we also showed that in the case of smallpox every day of delay in response increases the epidemic size exponentially. This was exactly what we saw with COVID-19. Both smallpox and the original SARS-CoV-2 had similar incubation periods and degrees of infectiousness. The cover-ups and delays in Wuhan resulted in global spread, by which time it was too late to put a lid on the genie's bottle.

In 2006, after being on a committee on stockpiling requirements for a biological event, I realised something was missing in our approach to risk analysis. The rest of the committee dismissed the risk of smallpox because it was eradicated, and they felt the probability of smallpox re-emerging was exceedingly low. The focus was mostly on anthrax, which is more accessible because it also occurs in nature. It was also used in the most recent attack at the time, the US anthrax letters sent in 2001 after 9/11. Yet anthrax is not transmissible from person to person. I felt a more layered risk analysis was needed to prioritise different pathogens that can be used for bioterrorism and set about doing one. These pathogens are classified as Category A, B and C. The Category A pathogens are the most serious, and include smallpox, anthrax, plague, tularaemia, botulism and viral haemorrhagic fevers (VHFs). VHFs include Ebola, Marburg virus and yellow fever.

In 2006, Mike agreed to be a co-author on the paper that arose from my risk analysis of Category A bioterrorism agents, which was published in the journal *Military Medicine*. I did a matrix of risk scores for a range of criteria such as infectiousness from human to human, stability in the environment, incidence of disease in a worst-case scenario, reports of genetic modification, global availability and ease of weaponisation. It involved a huge amount of reading and research to gather the data that was needed. The work showed that smallpox ranked second after anthrax and should not be dismissed as a serious threat. The next year I received a letter from the Association of Military Surgeons of the United States (AMSUS) to say I had won an unsolicited award for the work, the Sir Henry Wellcome Medal and Prize. That involved a trip to one of the most breathtakingly beautiful cities in the US, Salt Lake City in Utah, to receive the prize at the AMSUS conference and dinner. The experience inspired me to keep researching smallpox.

Preparing for smallpox

Smallpox was high on my radar again after the 2017 Evans horsepox experiment which proved that smallpox could be synthesised in a lab. It was the same year that monkeypox began causing large epidemics in Nigeria. These were the warning bells that should have prompted our government to prepare and stockpile newer drugs and vaccines. When monkeypox hit our shores, we had no antivirals or third generation vaccines.

Concerned about orthopoxviruses and wanting to raise awareness of the threat, I conceived and designed a tabletop exercise, initially called 'Exercise Mataika', to improve preparedness for the re-emergence of smallpox but also pandemics in general. In the simulation, an unknown epidemic begins in Fiji, and later spreads around the world through travel. My collaborator in the smallpox simulation was David Heslop, an academic at UNSW and an Australian Defence Force Army reservist and chem-bio-radiological-nuclear-explosives (CBRNE) expert. The simulation was first held in Sydney in 2018 with international and interdisciplinary participants including health, law enforcement and defence experts from 15 countries. Two attendees from the US military base at Pearl Harbor liked it so much, that we ran it again at Pearl Harbor in March 2019. Our friends at United States Indo-Pacific Command (US INDOPACOM) invited David and me to accompany them to Bangladesh in August 2019 to run the simulation again as part of a US–Bangladesh subject matter expert exchange. We tried to get our government interested in making it a trilateral event with Australia, with no luck. At the preparatory meeting in Dhaka, my name board read 'Raina MacIntyre, Team USA'. As an adjunct professor at Arizona State University, thanks to my home university UNSW's part in the PLuS Alliance with ASU and King's College London, I had been to the US five times in 2019, so I was quite happy to be on Team USA, given my own country wasn't interested. The US has always been special for me. I did a post-doc under a Harkness Fellowship at Johns Hopkins in the 1990s, and thoroughly enjoyed my time there.

We last ran the tabletop as 'Pacific Eclipse', a PLuS Alliance event on 9 December 2019 – the 40th anniversary of the eradication of smallpox. The event was held simultaneously in Washington, DC, Honolulu and Phoenix in a collaboration between the PLuS Alliance and the US INDOPACOM. We tailored the content for the US, including featuring the 2020 US presidential election. In the simulation, vaccine nationalism became part of an 'America First' campaign agenda, with no prospect of releasing American vaccine stockpiles to help less fortunate countries. Anti-vaxxers and disinformation spread on social media were part of the scenario too. As a result, because of national interest overriding global interest, disinformation and poor decisions, a pandemic that could have been stopped in less than a year became never-ending. Smallpox was well and truly back in the world, all demonstrated with mathematical modelling that we developed to underpin the exercise. We even simulated a world that changes forever because of the ravages of a pandemic, undoing decades of progress.

At the same time that we were running the simulation, and unbeknownst to the world, a real outbreak of COVID-19 had started in Wuhan. The severe impact of COVID-19 in the US in March 2020 and the country's poor response was a shock to most experts. The wealthiest country in the world was unable to even sort out testing for COVID-19, fumbling around with mistake after mistake while SARS-CoV-2 spread silently on the west and east coasts. From there they never really recovered. The Global Health Security Index was proven incorrect in ranking the US first in pandemic preparedness. It probably failed to consider three influential factors in pandemic control: leadership, culture and lack of universal healthcare. All of these came together to cause catastrophe in the US.

Public health measures for pandemic control are draconian by nature. To stop a virus spreading, we have to stop people having contact with each other, use masks, impose density limits, deploy social distancing and sometimes, when all else fails and the health system collapses, introduce lockdowns. During COVID, lockdowns bought time until vaccines were available. The culture matters because public health measures are less palatable in highly individualistic societies. Good leadership can overcome this by helping people come together on a difficult journey. But in 2020 the combination of highly individualistic culture and poor leadership was catastrophic for the US. Compounding this was the lack of universal healthcare. It meant that there were people who had to pay for a COVID-19 test and who were not able to access hospital care. In Australia, through 2020 and 2021, testing was widely accessible and free. But here, too, during the Omicron wave testing infrastructure collapsed. All of these lessons were exercised during the Pacific Eclipse smallpox simulation.

Explosion at Vector

The Vector Institute, Russia's main top security lab located in Siberia, and the US Centers for Disease Control and Prevention in Atlanta are the only two labs in the world housing live variola, the smallpox virus. In September 2019, an explosion occurred at Vector, shattering all the glass in the six-storey building. It was reported by news agencies and dismissed by the Russian government as inconsequential. Immediately, international virologists began minimising the incident. In major incidents like this one, health experts regularly choose to ignore science from fields outside their own and make up their own narrative from within their own narrow knowledge base. Here, they stated that any viruses released in the explosion would have been killed by the ensuing fire.

I, however, was concerned. I read up about the physics of explosions, and then discussed it with my engineering colleagues who did, in fact, have expertise and understood the physics of explosions. Together, we wrote an editorial in *Global Biosecurity* published in September 2019 in which we suggested that there was no scientific basis for denying that pathogens could not leak from the explosion. Explosions result in shock waves which are then followed by a fire. The detonation waves travel at far greater speeds than the flame, so that unburnt gas is propagated well ahead of the flame. Therefore, any viruses present and exposed through shattering of glass could have been accidentally leaked well ahead of the fire. We also speculated that the explosion may have been deliberate, to allow theft of bioweapons in the ensuing chaos. After all, this was Vector, the lab at the centre of the largest known bioweapons program in history during the Soviet era, Biopreparat, which experimented with engineering of smallpox to make it more contagious and deadly. Theft of smallpox and other deadly pathogens from labs like the Centers for Disease Control and Prevention or Vector is what intelligence and defence agencies worry about.

In the *Global Biosecurity* editorial, we stated that a disease with a long incubation period (10–12 days) may take several months to manifest with epidemic growth large enough to be detected. We advised that Russia and countries bordering Siberia, including China, Mongolia and Kazakhstan, should be monitored closely for serious emerging epidemics for at least the following three months (that is, through to December 2019). This all begs the question: is the COVID-19 pandemic, which began in late 2019 and had an initial incubation period of 12 days, a coincidence? I do not know the answer or if this could be connected, but back in September 2019 we were warning that a serious outbreak might occur in China. We also said that if theft of virus samples had occurred, an epidemic could occur anywhere in the world.

Publication of our paper caused a storm of hysteria. Our suggestions were loudly shouted down in a range of publications by the usual passengers on the biosecurity gravy train, who pop up regularly to silence any discussion of unnatural epidemics. Since 9/11 and the 2001 US anthrax attack, biosecurity has become big business, with a lot of people profiting from it. Some are recurrent apologists for the breaches and mistakes of nation states. The hysterical response to our paper about the Vector explosion was curious. Why were the stakes so high that they commandeered some high-profile publications as willing platforms to silence our paper, which appeared in a relatively unknown journal, *Global Biosecurity*, which is published by UNSW and of which I am an editor? These opponents criticised how quickly after the explosion the paper was published – well, my training in field epidemiology tells me that when it comes to transmissible pathogens you have to respond rapidly, not twiddle your thumbs for months. Academics who have never ventured outside of a university may pontificate ad nauseam before acting, and may want to pander to all parties, but anyone with field response training understands that epidemics grow exponentially, and time is of the essence. Perhaps the mob who shouted down the paper wanted a randomised controlled clinical trial, where one building containing viruses would be exploded deliberately and another not, with these compared for outcomes.

What we wrote, however, was better understood by law enforcement and intelligence types than the biosecurity experts, because of the perspective we provided. The situation offered a revealing case study of information warfare, something that features prominently in science and medical research. By this I mean a ping-pong game of narratives and counter-narratives – information warfare, a phenomenon seen recurrently in biosecurity. I suspect that our critics drew a lot more attention to our publication as a result of their loud and angry reaction than it may have otherwise received. And the furore of the reaction tells me that we probably had hit close to the mark in something we said.

8

INFORMATION WARFARE

ONE MORNING AN EMAIL LANDED IN MY INBOX which opened with: 'You continue this evil horseshit, prepare for the wrath of God ... good luck with that. I sincerely hope you repent. The truth is out there everywhere that these vaccines are POISON Russian roulette style. You promote it ... you're up for manslaughter ...' This is just one of the many crazy emails I regularly receive, amidst a steady trickle of threats and hate mail. Similar anti-vaccine propaganda is flooding social media and influencing many people around the world. Sometimes this is part of an organised disinformation campaign or even 'grey-zone warfare'. This is a military term referring to a rapidly proliferating type of warfare which falls outside of traditional definitions of war and peace. Instead of guns, bombs or open declaration of war, an array of unconventional but military aligned methods are mobilised to attack a country without provoking an actual military response. Grey-zone warfare can be used to subtly derail or destabilise a target country. This includes weaponisation of information, cyberwarfare as well as a range of more subtle attacks and threats such as 'lawfare' and economic manipulation.

Information warfare can include weaponised narrative, disinformation or harvesting of private information for psychological profiling. Information can be weaponised by nation states or other organised groups. In 2018 it was revealed that Cambridge Analytica, a British company, collected Facebook data from millions of users without their consent and provided profiling data to assist Republican candidates in the 2016 presidential election. They did this through an app, 'This Is Your Digital Life', which collected psychological profiling data on people who filled in their survey, but also allowed data harvesting from all their Facebook friends. It is estimated that fewer than 300 000 people used the app, but this enabled the data of over 80 million Facebook users to be harvested. There is also a suggestion that Cambridge Analytica was involved in the Brexit vote and outcome in the UK in 2016. All up, information warfare can be used to achieve particular strategic or political outcomes without the use of force.

Another example is the Russian use of disinformation to destabilise the US, both before and during the COVID-19 pandemic. In January 2021, infectious diseases expert Professor Peter Hotez published research in the scientific journal *PLOS Biology* titled 'Anti-science kills: From Soviet embrace of pseudoscience to accelerated attacks on US biomedicine'. In the paper, Hotez reflected on the poor US pandemic response and the influence of COVID-19 disinformation via social media. He outlined the mainstreaming of an anti-science agenda and the role of a hostile state, Russia, in amplifying this. The anti-science movement brought together disparate groups such as far-right extremists and alternative lifestyle groups with a common agenda (anti-vaccine, anti-mask, miracle cures), who were ripe for manipulation by organised disinformation campaigns. While Russia could not hope to defeat the US through military might, they have shown their mastery of information warfare, which has the ability to rock the foundations of democracy. Today technology exists enabling bad actors to use multiple grey-zone weapons simultaneously. Cyber-attacks can disable a power grid of a large city. Biological attacks by stealth, made to look like natural outbreaks, also are in the grey zone and can cause substantial disruption.

The big lie

The COVID-19 pandemic has illustrated the mainstreaming of disinformation, and the role of political leaders, doctors and other health experts as its willing mouthpieces. The following quote, often attributed (inaccurately) to Joseph Goebbels, points to the power of lies on a grand scale:

> If you tell a lie big enough and keep repeating it,
> people will eventually come to believe it. The lie can be
> maintained only for such time as the State can shield
> the people from the political, economic and/or military
> consequences of the lie. It thus becomes vitally important
> for the State to use all of its powers to repress dissent,
> for the truth is the mortal enemy of the lie, and thus by
> extension, the truth is the greatest enemy of the State.

In truth, Goebbels did not utter these words, but Adolf Hitler did touch on similar issues in his book *Mein Kampf.* Nazi propaganda did utilise the technique of what US analysts would later term 'the big lie'. American psychoanalyst Walter Langer, who compiled a psychological assessment of Hitler, wrote that

> His primary rules were: never allow the public to cool off; never admit a fault or wrong; never concede that there may be some good in your enemy; never leave room for alternatives; never accept blame; concentrate on one enemy at a time and blame him for everything that goes wrong; people will believe a big lie sooner than a little one; and if you repeat it frequently enough people will sooner or later believe it.

During COVID-19, the concepts of 'herd immunity' and 'endemic' have been weaponised as big lies. In 2020, despite being unscientific, the 'herd immunity by natural infection' theory became a household narrative. Experts, without even knowing what herd immunity is, nor the mathematical formula to calculate it, were at the podium spruiking it. The UK and Sweden led the way, and many countries followed. Promising herd immunity through mass infection of children, necessarily required a narrative that infection was harmless in kids. The herd immunity myth was a way of normalising infection and brainwashing people into accepting that getting infected (and government doing little to offer protection), is good for them. The same spin-masters also fabricated a brand new concept, 'immunity debt' which means preventing infection is bad, because ultimately you will get infected anyway, so you may as well get infected now. This resulted in the dystopian twin propaganda of 'get infected to prevent infection' (herd immunity) and 'preventing infection is not worth it' (immunity debt) to pave the way for mass infection.

As a concept, herd immunity arose out of vaccine programs. No infection has ever controlled itself without the use of vaccines. In the pre-vaccine era, smallpox, for instance, caused recurrent, large-scale cycling epidemics for centuries. The same was true of measles and the patterns of SARS-CoV-2 to date have been the same. And the idea of 'immunity debt' goes against the principles of public health. Would we say to drivers, 'get drunk and forget your seatbelt today to get your road accident over and done with'? Or to young people, 'smoke as many of your cigarettes now, to get your coronary bypass earlier and get it over with'? That is essentially what this propaganda is saying about COVID but couched in pseudoscience.

In everyday usage, the word 'epidemic' is widely applied, used to refer to anything to create a sense of urgency: An epidemic of ice. An epidemic of diabetes. An epidemic of burglaries. True epidemic infections are distinct from endemic diseases such as malaria or diabetes. An endemic disease may exist in high or low numbers but does not change rapidly over time. Changes, if any, are slow and occur over years. Endemic diseases do not crash the health system and decimate your workforce in a short period of time. They do not cause supermarket shelves to be empty. In contrast, epidemic diseases can grow exponentially over days or weeks. An epidemic which spreads globally is a pandemic. Epidemic and pandemic diseases require immediate surge capacity in health systems – we saw this during the SARS-CoV-2 pandemic, which is a typical epidemic infection. Globally, we saw health systems overwhelmed over very short time periods.

Contrary to what the term might suggest, elimination does not mean absence of epidemics. It means herd immunity has been achieved, and therefore any epidemic that starts will not be sustainable in the long run because too many people are immune in the population. Immunity may result from infection, or by vaccination. Some infections, like measles, give lifelong, strong immunity, and others like SARS-CoV-2 give temporary immunity that wanes and is not as strong as vaccine immunity. The required herd immunity to eliminate an infection can be calculated mathematically using the known R0 value. The required proportion of the population that must be immune to achieve herd immunity (H) is not an arbitrary number that can be made up – it is calculated by the formula $H=1-(1/R0)$. Eradication is when a disease is absent globally. Elimination applies to regions or countries and is usually a goal in the context of a WHO eradication effort. Countries obtain certification of elimination by WHO. Australia achieved WHO elimination status for measles, but we still see occasional outbreaks of measles, usually imported through returning travellers. However, they never become sustained in the same way SARS-CoV-2 is.

When SARS-CoV-2 vaccines became available, not one of the experts who excitedly advocated infection as the way to prevent further infection was interested in aiming for herd immunity through vaccines. Formally, when elimination or eradication are not possible, 'control' is meant to be the aim. During the COVID-19 pandemic, we saw experts who lacked the requisite knowledge of these differences, fabricate an entirely new term: 'suppression'. They then proceeded to create a polarising narrative about elimination versus suppression and demonised any experts who were pro-elimination.

In 2022 we then moved into an entirely new phase of disinformation, where we were told by experts that the pandemic was over, and we could reclaim the lives we had left behind in 2019. Yet SARS-CoV-2 has not mutated into a cold, as some experts predicted, but is still crushing health systems and killing people. In Australia, the Omicron strain caused more deaths than all other waves and became the third leading cause of death in 2022. And yet the narrative on deaths was spun and twisted to make us accept that hundreds of deaths a week is perfectly normal and unavoidable. It is not. We could reduce these deaths, hospitalisations and long COVID by using a vaccine-plus strategy of layered prevention including boosters, masks, testing and tracing. Instead, these measures have been abandoned and very little effort has been made to get third-dose vaccination rates above 90 per cent. Those of us who speak up and advocate for prevention of illness and death in the population are called extremists, aligning us with terrorists and other dangerous groups. Other peddlers of disinformation and mass infection proclaimed themselves to be centrists, in a further effort to legitimise themselves and to marginalise critics like me. The pandemic has effectively seen public health become public disease, with doctors and experts assuring us that mass infection is a good thing. We are even being urged to actively seek to become infected in order to avoid infection. This ridiculous proposition is made even more so by the fact that one infection does not protect well against subsequent infections.

Many believe that science is unbiased. But just ask anyone who has worked in research or knows the history of tobacco science or climate science – the same vested interests and networks of power exist in science as they do in any field. In epidemiology, anyone with a hypothesis can design and conduct a study to bolster that hypothesis. A whole industry in counter-narrative research and biased Cochrane reviews, supposedly the gold standard of medical evidence, can be generated to support powerful vested interests.

Cigarettes and stomach ulcers

Instances of doctors, scientists and medical experts actively spreading disinformation to oppose the truth are as old as recorded history and should not surprise anyone. Doctors generally enjoy a high degree of trust among the population, which makes them a valuable commodity for promoting vested interests such as those of corporations, nation states or other powerful actors. Many physicians choose to prioritise personal profit over the best interests of the public. The best example of this is cigarette smoking. Due to concerted efforts by the tobacco industry to fund and publicise counter-narrative research by willing doctors, it took over 100 years to acknowledge the harms of smoking. Initially, doctors who tried to sound the alarm about the dangers of smoking were labelled 'alarmist' and 'quacks' (in today's language they would be called 'extremists', while the pro-smoking doctors would be called 'centrists'). Other doctors appeared in advertising campaigns for cigarettes, and conducted research funded by the tobacco industry to show favourable results. Many top scientific journals colluded and willingly published these results, designed to confuse and confound the growing evidence that smoking was harmful. Cigarette brand Camel used the line 'More doctors smoke Camel than any other cigarette' in their advertising. Until the 1950s, tobacco advertising even appeared on the pages of top medical journals.

A more recent example has been the reaction to findings of the cause of stomach ulcers. Through their research, Australian professors Barry Marshall and Robin Warren, both based in Perth, discovered that the bacteria *Helicobacter pylori* was the cause of stomach ulcers. Warren first noticed the bacteria under a microscope, and Marshall performed a simple but brilliant experiment by drinking the bacteria and then proving it causes stomach inflammation, which could be cured by a course of antibiotics. Their discovery earned them the 2005 Nobel Prize in Physiology or Medicine, but it also resulted in enormous pushback from the medical-industrial complex, with top medical journals more than willing to silence them. When the medical research community does not want a new finding to be published, journals, editors, reviewers and the whole tight-knit, powerful medical publishing infrastructure will close ranks and not only suppress truth but publish counter-narratives to cast shade on it.

For mainstream gastroenterology, Marshall and Warren's finding was blasphemous. The prevailing theory at that time was that 'stress' causes stomach ulcers, and there was a belief that the stomach was sterile, and no bacteria could grow there because of the acid. The drugs Tagamet and Zantac, H2 blockers (which neutralise stomach acid), were the first of any drugs to each break the billion-dollar sales ceiling. These drugs represented a huge business opportunity as rather than requiring a single course of antibiotics to cure the ulcers, patients needed lifelong treatment with a H2 blocker. That was a lot of profit to be lost. Marshall and Warren's findings threatened profits from antacid drugs as well as the profits made by surgeons from ulcer surgery and psychiatrists from treating 'stress'. Since stomach ulcers can also become cancerous, the breakthrough by Marshall and Warren not only resulted in a cure, but also a dramatic reduction in stomach cancer rates.

Marshall and Warren had great difficulty getting the research published, rejected over and over again from major medical journals and conferences. Even the Gastroenterological Society of Australia rejected their work for their annual conference in 1983. When it eventually did get. published, they were subjected to a flood of ridicule and counter-narrative research. It took more than 20 years for their landmark work to be recognised by the awarding of the Nobel Prize for Physiology or Medicine. Even then they were subjected to sour and dismissive comments from members of the Australian medical community. The experts who blocked Marshall and Warren's work had probably retired by then, still entrenched in their outdated views. A quote from the 1918 Nobel Laureate in Physics, Max Planck, summarises for me the problem of ideology in science and resistance to change: 'A new scientific truth does not triumph by convincing its opponents and making them see the light, but rather because its opponents eventually die, and a new generation grows up that is familiar with it.'

In historical examples and during the COVID-19 pandemic, we have seen scientists opposing accountability for their own peers, and scientific journals willing to publish counter-narratives to protect the interests of scientists. Powerful lobby groups and vested interests regularly control the narrative in science. They can gatekeep what is published in journals and who sits on committees, they can influence input into policy and throw everything at researchers whose findings are inconvenient or opposed to a dominant ideology. In 2020, at the start of the pandemic, several top journals published early opinion pieces insisting that the virus has a natural origin and attacking anyone who suggested otherwise as a conspiracy theorist. Every time more evidence emerged that shed light on the lab leak theory, scientific journals published new counter-narratives to peddle the natural origin theory. There is a serious imbalance in scientific journals with barely any published evidence supporting a lab leak, while a huge volume of opinions and inconclusive research supports a natural origin. And yet there are findings that do suggest that the COVID-19 outbreak was not natural. In Australia, for example, Professor Nikolai Petrovsky, discovered quite by accident, while working on developing a vaccine against the virus, a signal that SARS-CoV-2 may be engineered but had difficulty getting it published. The Sverdlovsk anthrax accident in 1979 is another example. In that instance journals supported experts in their efforts to first deny a lab accident and later, after Boris Yeltsin confessed, to minimise the possibility of offensive bioweapons development in the USSR.

Hygiene theatre

Most of 2020 was spent emphasising the importance of hygiene theatre, 'deep cleaning' (a reassuring term that sounds very impressive) and scrubbing of hands. In India, international cricket legend Sachin Tendulkar was featured in a health promotion ad to show people how to wash their hands while singing happy birthday twice. Similar advice was promoted elsewhere. In the end the whole world was washing their hands until the skin peeled off, showing how effective health promotion can be when it is done with purpose and commitment. Meanwhile, however, no such campaign was used to convey the dangers of airborne transmission. Even today most people still are unaware that SARS-CoV-2 is mainly caught through inhalation. Remarkably, early in the pandemic, the WHO, the US Centers for Disease Control and Prevention and most countries even adopted an anti-mask stance. The message was 'Do NOT wear a mask. It might be dangerous. Just wash your hands.' It was also suggested that a mask was only needed when you had symptoms of SARS-CoV-2 – implying that it was fine for an infected person to go out wearing a surgical mask. Asymptomatic infection meanwhile was completely ignored. And in fact, until June 2020 the WHO maintained that there was no significant asymptomatic transmission. In reality, the most infectious period is in the one to two days before the onset of symptoms, and 30–50 per cent of people never develop symptoms but are still contagious all the same.

One educational cartoon on COVID-19 in early 2020 featured an influential figure stepping into an elevator with two masked people and telling them that they should be unmasked like him. When the air you breathe is the most important determinant of transmission of SARS-CoV-2, an elevator of all places is the most risky – a small, enclosed space with no ventilation. People in my own family and circle of friends understand airborne transmission because I have drilled it into them. Recently one person, who had already brought a portable air purifier with a HEPA filter into his office, made me laugh when he told me he goes up in the elevator more than 20 floors in his city office block while holding his breath. I told him that a good quality N95 would be a better option. If I can teach those close to me the requisite knowledge on how to protect themselves against COVID, surely a well-thought-out media campaign could similarly empower people to improve their chances. It's the air you breathe that matters. Open a window. Wear an N95. Use an air purifier. Queensland University of Technology made a great video on safe indoor air featuring Professor Lidia Morawska, called 'Ventilation Revolution', and then the advocacy group The John Snow Project made another. However, without the power of a government campaign behind it, it has not reached the masses, who continue scrubbing their hands.

I did not know Professor Morawska until the COVID-19 pandemic, but we quickly connected over shared concerns about airborne transmission. A physicist and world-leading expert on safe indoor air, Professor Morawska is, like many important scientists, largely ignored in her own country and has not been given a seat at the table to inform Australia's pandemic response. Infectious diseases doctors lack knowledge on principles of airflow, engineering of buildings and movement of aerosols – all important determinants of infection risk. When dealing with an airborne infection, we need a very broad range of experts at the table, not just those from health disciplines. We need engineers, architects, occupational hygienists and a range of other experts. Reading Professor Morawska's research and that of other experts taught me more about safe indoor air and ventilation than I had ever learnt in my career. In 2021 Professor Morawska was even named one of *Time Magazine*'s 100 most influential people for her role in driving change in the WHO around airborne transmission of SARS-CoV-2. She led a letter signed by over 200 international scientists challenging WHO's position of denial of airborne transmission of SARS-CoV-2.

Our own research played a small role in this turnaround. In 2018, I had established an aerosol lab and collaborations with engineering experts at UNSW. Together we work as an interdisciplinary team of engineers, scientists and medical researchers on mask research, ranging from understanding the behaviour of human respiratory aerosols to developing novel fabrics for a re-usable mask. In April 2020, we published a paper in the *Journal of Infectious Diseases* showing that the '1–2m rule' of social distancing, widely promoted during the pandemic, is not supported by science. This rule is used by infection control experts to justify surgical masks and handwashing as adequate protection (as long as you keep your distance). We showed that large droplets can travel further than 2 metres, and that social distancing policy was based on entrenched ideology rather than evidence. It was an influential and highly cited piece of work, led by our engineering PhD student Prateek Bahl, and among the first to challenge the denial of airborne transmission and N95s for health workers.

Politics and power, not public health

During the COVID-19 pandemic, disinformation has been repeated so many times by experts and politicians alike that it is believed by the majority of people. People have been told the pandemic is over and have celebrated that fact, while around them schools are barely functional due to recurrent outbreaks, airports are unable to operate because of mass absenteeism, and people needing coronary bypass surgery cannot get it because the health system is overloaded. As I write, in Australia the number of deaths from COVID-19 are equivalent to a 737 aeroplane crashing once a week – if even one plane crashes or a single person is taken by a shark, it is front-page news, but this ongoing, relentless parade of death passes without comment. Then there is the obfuscation of data on deaths, which enables the cheerful spin to be continued, and worse, some deaths to be minimised as trivial or even unrelated to COVID.

In the US, we have seen the dire outcomes of poor leadership, with leaders peddling unscientific theories and lethal 'miracle cures', and medical experts actively discouraging public health interventions such as the wearing of masks. Basic public health measures such as masks and vaccines have become politicised and seen as symbols of oppression in many countries. In 2022 the director of the US Centers for Disease Control and Prevention (CDC) referred to masks as a 'scarlet letter', a symbol of shame. Throughout the pandemic, the CDC has swung wildly between advising against masks to recommending N95 respirators for the community, with multiple flip-flops along the way, despite many studies confirming the effectiveness of masks against COVID-19. The lasting damage and mistrust this will cause cannot be underestimated.

During the COVID-19 pandemic government advisors have often been chosen on the basis of their willingness to follow the political line, rather than expertise. Being a mouthpiece for power comes with rewards, whilst speaking truth is punished. Almost everyone I know who has spoken up for prevention of illness and death has paid a price. Some in government jobs have lost their positions. Others have been reported to the medical regulators or had other attacks against them. I have come to know many good people through the pandemic, some quiet achievers who work behind the scenes, and some who are vocal on social media about protecting the health and welfare of people. It is the vocal ones, all good people who were driven by a desire to protect others, who have paid the highest price.

In 2020, I was on a large government committee for COVID research advice, probably invited on to keep me in line. Frustrated by the lack of good policy and failure to consult a broad range of experts, in September 2021 a group of experts who were deeply concerned about the worsening COVID situation formed OzSAGE. OzSAGE is an independent voice a bit like the UK's Independent SAGE. The three of us who were also on the government committee were called by the deputy Chief Medical Officer to relay the displeasure of the government about OzSAGE. We were asked to choose between staying on their committee and being on OzSAGE. I told him OzSAGE could be an additional resource for government, outlined the talent and multidisciplinary expertise we could offer, and our willingness to work with them, but to no avail. OzSAGE was perceived with hostility. All of us stepped off that committee and chose OzSAGE, which has gone on to do important work, issue a series of valuable advice documents and hold federal and state governments to account. When so many doctors and other experts have switched from public health to public disease, OzSAGE has shown me hope, that there are still plenty of experts with integrity who care about prevention of disease and protection of vulnerable members of society.

Useful idiots

Disinformation has a more sinister underbelly than simply personal greed of experts. Some experts are useful accomplices to larger agendas and backed by well-resourced organisations or entities. The Great Barrington Declaration (GBD) arose in 2020 out of a meeting of experts convened by the American Institute for Economic Research (AIER) in Great Barrington, Massachusetts. The GBD promoted herd immunity by mass infection, while providing 'focused protection' (which essentially amounted to being excluded from society) to the elderly and vulnerable. Even then it was known that immunity from infection and vaccines wanes. The GBD falsely stated that COVID-19 was less dangerous than influenza to children.

Investigative journalist Nafeez Ahmed from *Byline Times* has done some exposés of the GBD's disinformation campaigns during the pandemic. He points out the climate change denial agenda of the AIER, which receives funding from an oil billionaire. AIER itself owns an investment firm with holdings in tobacco, fast food, energy, oil and gas companies.

The GBD legitimised a movement which was strongest in countries like the UK and Sweden. Both promoted widespread infection and used experts to propagate the message. In the case of the UK, children were denied vaccinations long after peer countries had finished vaccinating them, because of the hope that infecting children would help achieve herd immunity. In Australia a group of experts were aligned with the GBD and also campaigned against vaccination of children. They repeated the same message, that less than 1 per cent of kids have complications. Well, that is also true of most diseases we vaccinate children against and is a standard argument of the anti-vaccination lobby. The statement that COVID is mild in most children is true. But is this an argument to warrant denying vaccination? Like COVID, polio is mild or asymptomatic in 99 per cent of children. Only 1 per cent of children become paralysed, and yet we consider this unacceptable. In Australia in 1951, at the peak of polio, there were 357 deaths. Now, 70 years later, we still protect children against polio. From 1956 to 1975 we had 356 deaths from measles in Australia – less than 20 deaths a year. And yet we think it's worth protecting children from measles. With whooping cough, it was 290 deaths a year between 1910 and 1942. In the US, in one of the worst recent influenza years, 2017–18, 186 children died of influenza. During the COVID pandemic over 700 children died in the first 18 months.

To be consistent with the high value we place on protecting the health of children against other infections, for children who are going back to school and unvaccinated, we should do all we can to protect them. That means ensuring they have safe indoor air and widespread accessibility to rapid point-of-care testing, that parents and teachers are vaccinated and masked (along with the children), and that classes are generally designed in such a way as to reduce risk of infection. Yes, some children will have trouble wearing a mask, but most won't. The same people who campaigned against vaccines for children also campaigned ruthlessly against masks and safe indoor air – which makes you wonder how sinister their intent really was. Infection in a child may harm the child, but also adults around them, some of who may be medically vulnerable. In the US alone, 200 000 children have been orphaned by COVID-19. Globally, an estimated 5 million kids have been orphaned by COVID-19.

In the UK a theatrical narrative of 'mystery' was spun in April 2022 in response to an epidemic of hepatitis in children. The theatrics spread widely when other countries started observing the same thing. If COVID is so trivial in kids that you don't even need to vaccinate them, what happens when serious complications become apparent? When experts trivialise illness in children and deny them vaccination, that necessitates a 'mystery' to explain a serious complication. It's called escalation of commitment to a failing proposition – a well-known human behavioural phenomenon, taught in business schools around the world. When people become invested in a certain position and evidence accumulates that their position is wrong, the normal human response is not to back down and admit wrong, but to dig in and become even more fanatical about defending that position. This is why, for instance, the Space Shuttle *Challenger* exploded in 1986, killing all on board. The engineers warned NASA management that if *Challenger* was launched under the prevailing temperature conditions, there was a 100 per cent probability it would explode. Management, however, were so invested in the launch as a means to gaining more funding for the agency, they minimised the warnings of the engineers, ridiculed and sidelined them and went ahead. The price paid was in human lives and ultimately, loss of the very thing that was driving management: money and profit.

We know that multisystem inflammatory syndrome, a rare complication of COVID in kids, commonly includes hepatitis. When COVID-19 rates are extremely high, even rare complications will occur in numbers large enough to be noticed. Although some other cause may be identified, the most likely explanation is COVID multisystem inflammatory syndrome. We also know that this complication is not immediate, but occurs weeks after the initial infection, by which time many kids will test negative to the virus. Even serology may be negative, because kids do not develop antibodies to SARS-CoV-2 as reliably as adults do. Two viruses have been found in kids with hepatitis, one being SARS-CoV-2 and the other being adenovirus, but adenovirus is not found in the liver of affected kids. The pontificators have rushed to blame adenovirus, which is often a passenger found in kids with other illnesses and has never caused severe hepatitis before in healthy kids. The epidemic of hepatitis occurred during the Omicron pandemic wave. Studies show substantial liver involvement in

SARS-CoV-2 infection, so the biological plausibility that it is SARS-CoV-2 related, is strong. It is possible that the Omicron variant is more likely to cause hepatitis in kids, or simply that the infection rates are so astronomically high that a rare complication is being noticed. The doctors who campaigned viciously against vaccinating children are still creating theatrics around hepatitis in kids being a 'great mystery'. They are desperate to find any explanation other than COVID, because they are digging in to defend their position that COVID is trivial in kids. How could doctors knowingly engage in such deception around the lives of children, you may ask.

9

TRUST ME, I'M A DOCTOR

THE MEDICAL PROFESSION AND MEDICAL RESEARCHERS are trusted. Numerous studies show that most people trust doctors, so they find it difficult to believe that doctors could advocate for harm or, worse, be involved in deadly research designed to kill. They assume good character is conferred by a medical degree, but the involvement of doctors in biowarfare research, torture and murder says otherwise. Some cases are individual examples of doctors who kill and harm, such as Harold Shipman, a British GP who killed more than 15 of his patients, or Michael Swango who killed at least 60 patients. Several nurses have also been caught, including German nurse Niels Hoegel, who injected patients with cardioactive drugs to make them have a cardiac arrest, and then played hero in resuscitating them. He was convicted of 85 deaths from 1999 to 2005, but the true number may have been much higher. These cases simply illustrate that the whole spectrum of human character is represented in medicine – including psychopaths.

Dual loyalty

Sometimes the medical profession as a collective carries out mass killings and other atrocities. Dehumanising of 'the other' is central to the collective medical profession committing atrocities against humanity. In a clear breach of medical ethics and the Geneva Convention, doctors can often act as agents of the state, giving it primacy over the patient. After the horrors of Nazi medicine were exposed, the Geneva Convention of 1949 states that medical personnel 'shall not be compelled to perform or carry out work contrary to the rules of medical ethics'. Dual loyalty is when a third party, usually the state, competes for a doctor's loyalty to their patients. When doctors allow state interests to override their own duty of care to patients or public health, breaches of medical ethics can occur. There are numerous examples of this historically and in contemporary times, including in Nazi Germany, Apartheid South Africa, and the War on Terror. It is much harder to see and accept contemporary examples. When doctors have skin in the game and are acting against the interests of patients, they may reject examples from the past as aberrations.

The atrocities committed by doctors in Nazi Germany are well documented. The executive director of the Hartmannbund, one of Germany's largest medical associations, stated in 1933 that 'To serve this State must be the sole objective of the medical profession.' As a result of German soldiers becoming traumatised by carrying out mass executions of civilians, the Nazis later developed mechanised mass killing to rid the soldiers of the burden and allow them to carry on the war. Significantly it was Nazi doctors who enthusiastically came up with the

Zyklon B gas chambers, actually exceeding the demands of the state by taking the initiative to design the apparatus for mass murder. Many apologists claim that Nazi doctors were coerced or were the victims of a 'slippery slope'. Yet in total contrast to their German counterparts, Dutch doctors in Nazi-occupied Holland, refused en masse to obey orders for forced euthanasia and sterilisation, despite the orders being framed in a utilitarian context, rather than as genocide. Furthermore, if the Nazi doctors had been truly coerced, some sort of crisis of conscience would remain in the historical record – and yet there is none, as noted by Nobel Laureate and concentration camp survivor Elie Wiesel. The Nazis viewed the terminally ill, disabled or ethnically targeted people as *Untermensch* (subhuman) and an economic burden. The role of doctors in the Holocaust proves how far doctors can stray when they fail to consider the basic question of 'who do I serve?'.

Doctors who collaborated with security forces in South Africa were similarly not coerced but identified closely with the state's ideology. This made involvement in military offensives against the Black Liberation movement and developing biological and chemical weapons acceptable to them. The death of South African activist Bantu Stephen Biko in 1977 at the hands of security police is a clear example of medical collusion. The doctors summoned to treat Biko in police detention ignored evidence of injury, including signs of brain damage. Even after he was found to have impaired speech and gait and later exhibited 'the Babinski sign' (a sign of significant brain damage), an examining doctor certified that he 'found no evidence of any abnormality or pathology', thereby falsifying the medical record to hide the use of torture. Biko received no treatment and doctors allowed him to be transported to a prison hospital lying on the floor of a van in a pool of his own urine.

Some decades ago, I was given a biography of Biko. At the back of the book was printed his autopsy report which showed, among other fatal injuries (including three brain lesions from blunt force), that the force used on Biko caused his thoracic aorta to rupture. The thoracic aorta sits deep in the chest cavity against the front of the spine and is heavily protected by other organs and the chest wall in front of it, and by the spine and chest wall behind it. It is so deep inside that to rupture it would require unimaginable force. As a young doctor at the time, it was horrifying to read and to understand that other doctors could be capable of falsifying such atrocities. The state-controlled South African Medical and Dental Council ruled that there was no evidence of medical misconduct, despite vast evidence to the contrary, and even obstructed later attempts by individual doctors to reopen the investigation. Much later, the Council acknowledged that it failed to act appropriately because patient welfare was seen as irrelevant when applied to 'the enemy'.

When doctors succumb to a political ideology or climate of fear, they may choose to violate the rights of human beings that they see as 'the enemy'. A doctor may also breach their duty of care when they no longer view the person as their patient, or even a human being. Doctors directly employed by the state may face even greater pressure to meet the demands of the state and their hospital, thus compromising their independence. In Rhodesia, in a secret detention centre in Mount Darwin, doctors tested a range of biological and chemical weapons on captured human subjects, testing different doses and combinations of deadly agents. Dead bodies were thrown down mine shafts to get rid of the evidence, which was unearthed years later when mass graves were discovered.

The involvement of medical professionals in the Australian government's treatment of refugees is another case in point. Australia's policy of mandatory detention of asylum seekers has continued unabated since 1992 through both Liberal and Labor governments. The politicisation and demonisation of asylum seekers in the community and media has resulted in medical personnel colluding with inhumane actions. While detention centres were wholly onshore in Australia, there was continual scrutiny by community groups, until detention centres moved offshore under the Gillard government in 2012. Later, doctors were also criminalised under the *Border Force Act 2015* by the Australian government, facing up to two years in prison if they spoke out about child abuse or neglect of asylum seekers in detention. In 2016, Doctors 4 Refugees launched a High Court challenge to the constitutional validity of secrecy provisions within the *Border Force Act* and had this overturned by October of that year. After 15 years of offshore detention, the Medevac bill passed in 2019, through the advocacy of medical MPs like Dr Kerryn Phelps, finally allowing sick asylum seekers to access healthcare in Australia, only to be repealed shortly afterward. The offshore indefinite detention policy continues, with doctors caught up in dual-loyalty situations.

Silencing of dissent

During the COVID-19 pandemic another example of dual loyalty has become apparent, as medical practitioners in Australia are prevented from speaking publicly about the national vaccination program, except when disseminating the official government messaging. To silence doctors, a two-pronged strategy appears to have been used, utilising legislation around the Therapeutic Goods Association (TGA) as well as the regulatory body the Australian Health Practitioner Regulation Authority (AHPRA). The TGA legislation is designed to prevent doctors from advertising drugs or vaccines. AHPRA's core remit is 'to ensure the community has access to a safe health workforce across all professions registered under the National Registration and Accreditation Scheme'.

This situation arose in early 2021 during the rollout of the AstraZeneca vaccine when its association with a serious and potentially fatal clotting syndrome called thrombosis with thrombocytopenia syndrome (TTS) became clear, especially in younger adults. When young people died of TTS, many doctors expressed legitimate disquiet over the use of the vaccine on younger people. There were also concerns about over-reliance on this vaccine and shortages of the mRNA vaccines, which had higher efficacy based on clinical trial data. A poor procurement strategy by the Australian government failed to diversify vaccine supplies and relied heavily on a local agreement to manufacture the AstraZeneca vaccine domestically. The government over-invested in the AstraZeneca vaccine and, with few other options, silenced doctors using the threat of imprisonment and deregistration. This opened the way for lay people and non-clinician epidemiologists with no knowledge of the complexities of clinical vaccine-adverse events free to say anything they liked publicly (which they often did, waxing lyrical in the media about vaccine-adverse events from a position of utter ignorance). Meanwhile doctors with genuine concern for patients were gagged.

Doctors who suggested one vaccine is superior or safer than another risked prosecution and up to one year in prison from the TGA and disciplining and deregistration by the AHPRA. A doctor who has valid concerns about the vaccination program is subjugated to the power of the state and prevented from speaking out for the public good. At the same time, a whole army of doctors became slavish cheerleaders for the government and began policing and attacking those of us who stuck to the science and spoke in the best interests of the people. In 2022 Western Australian rural doctor David Berger was disciplined by the AHPRA for language used in public health advocacy and ordered to undertake 're-education'. Many other doctors I know who dared to voice their concerns for patients were reported to the AHPRA by other doctors.

In June 2022, the Medical Council of NSW tightened their policy, stating that 'Inappropriate social media activity includes: Activity that contradicts public health orders, public health messaging or reputable scientific evidence such as: Making comments, endorsing or sharing information, or posting "likes" or "dislikes" with or without additional comments.' If that does not chill you, it should. It is so broad that it theoretically allows them to target some doctors while letting others get away with a breach. A doctor protesting the repeal of the

Medevac law would be going against official government policy and could be in breach of this policy. In the future, a doctor would not be able to protest a hypothetical state-sanctioned eugenics policy. In other words, the policy subjugates doctors to obedience to the state and does not allow protest or dissent that may be in the best interests of patients. Other defined inappropriate conduct includes 'Unprofessional, disrespectful or threatening communications with or about patients and/ or other practitioners, including through closed group social media channels.' I have certainly had plenty of disrespectful and inappropriate comments made about me or amplified by other doctors, but I am not aware of anyone facing repercussions for it – whilst good doctors I know have been punished for advocating for better public health.

I myself was attacked for a factual statement in a media interview, where I referred to 'the less effective AstraZeneca vaccine'. After TGA and AHPRA silenced doctors in early 2021, I did not comment on the vaccines at all. Despite all the spin, propaganda and crushing of dissent by the government and their favoured doctors, what I said was a scientific fact, borne out in randomised clinical trials, real world studies, and evidence of reinfection with new variants – the AstraZeneca vaccine has lower protective efficacy than mRNA vaccines, which was the other type of vaccine available in Australia at the time. One doctor tweeted that he would be complaining about me to the vice chancellor of my university and to my immediate boss for making that factual statement. His letter of complaint was so bizarre it even suggested that I had caused vaccine hesitancy in Australia, and essentially asked how the university would punish me. I am quite used to being bullied at work by white men. I have lost count of the number of emails I have received by men seeking to take something from me or bring me down, who, when facing my pushback, send a threat and copy in my boss or other powerful men to intimidate me. Suffice to say I was not disciplined as this man had hoped. Everyone was aware that ATAGI flip-flopping every few weeks about the AstraZeneca vaccine, media coverage of the vaccine-related deaths of several Australians from TTS, lack of trust in government by the community, and various other factors were responsible for vaccine hesitancy and that it was rather fanciful to suggest it was my personal doing. In another battle, a group of doctors campaigned against COVID-19 vaccination for children, insisting it was a trivial disease and that the vaccine was more risky than COVID-19. They were following the UK model, which was to deny vaccination to children long after other countries were protecting their children. There is published research showing an increased risk of several complications in children, even after mild COVID, but this did not stop them.

I have worked in vaccinology for 30 years and have published hundreds of research papers on vaccines. In my 15 years at the National Centre for Immunisation Research and Surveillance, I had to deal with the anti-vaccine lobby on a regular basis. These arguments generally run like this: infection is mild in most kids; building up natural immunity is good; vaccines are risky. It was shocking to witness that the arguments of the paediatricians who campaigned against vaccinating kids were the exact same arguments of the anti-vaccine lobby. When I tweeted the scientific truth that polio was mild in 99 per cent of people who got infected and only caused paralysis in 1 per cent, this same group became outraged, demanding I take down the tweet. The attacks were so ferocious and threatening that I did take down the tweet, wondering if the truth was just too confronting for these people to see their complicity in causing mass harm to children. Yet all the infections we vaccinate children against – measles, chickenpox, hepatitis B, pertussis – have death rates in children that are similar or even lower, than COVID-19. In the UK, a deliberate decision to allow mass infection of children appears to have been made without any consideration of the long-term effects on their development or of the parity with other vaccines that are routine for children. Somehow the information warfare that had split the US into red and blue and brought about Brexit in the UK, had permeated into the minds and hearts of previously reasonable doctors in Australia. Fortunately, ATAGI recommended vaccination of children in Australia and these same paediatricians swayed with the breeze and pretended they supported it all along, but continued to run counter-narratives about COVID-19 being trivial in kids.

Whistleblowing

Whistleblowing usually occurs as a last desperate act following a series of failures. The most famous medical case of whistleblowing was the Bristol Royal Infirmary (BRI) in the UK, a 400-bed, tertiary referral hospital serving a population of 7 million in the South West of England and South Wales. Dr Stephen Bolsin, an anaesthetist, joined the BRI in 1988, after completing his training at the Brompton Hospital in

London, a national referral centre for adult and paediatric cardiothoracic surgery. The BRI had two cardiothoracic surgeons performing surgery on adults and children. Bolsin noticed that operations at BRI seemed to take longer than operations at Brompton, and patients (both adult and children) seemed to be suffering more complications and dying more frequently. One surgeon took three times as long as average. Dr Bolsin began to collect data and keep records of cases and outcomes. In 1990 Dr Bolsin had enough data that something was amiss at BRI and tried to raise his concerns. He was met by resistance from cardiac, surgical and even some anaesthetic colleagues and was discouraged from keeping records. In the same year the BRI became part of United Bristol Healthcare Trust, one of the UK National Health Service's first and largest hospital trusts. Dr Bolsin wrote to the CEO of the trust about his concerns. No action resulted.

In 1992 Dr Bolsin decided to leave BRI. When he stated his reasons, his director asked him to collect 'hard evidence' so that he could deal with the matter. He began collecting data to compare with national register data in a project funded by the Department of Health. By 1993 he stopped doing paediatric cases and only worked on adult cases. Dr James Wisheart (one of the surgeons in question) became Medical Director of the Trust, thus being medical advisor to the CEO. Bolsin presented completed data to the CEO, Director of Anaesthetics and newly appointed professor of cardiac surgery. The data showed a shockingly high rate of deaths at BRI, including in children, operated on by these two surgeons compared to the national average. Another professor of surgery had already approached the Department of Health with the same concern. The

Department sat on a report highlighting poor performance in cardiac surgery in BRI, and the surgeons kept operating. More children died.

In June 1994 a new paediatric surgeon began working at BRI, and the professor of cardiac surgery believed he had made an agreement with the incumbent surgeons that no further arterio-venous switch operations were to be done by them. Yet in December a routine operation list showed that Dr Janardan Dhasmana, the other surgeon in question, was going to do the procedure he had been expected not to do, on an 18-month-old child. Fearing for the life of the baby, Dr Bolsin and the professor of cardiac surgery notified all clinical and non-clinical managers within the hospital, but no one was prepared to intervene. They then contacted the Department of Health, who contacted the CEO and requested the surgery be postponed or the child be transferred – the CEO did not act. An emergency meeting held the night before the operation showed 69 per cent mortality in babies operated on at BRI by these two surgeons compared to a UK average mortality of 6.5 per cent. Everyone at the meeting (except Dr Bolsin) agreed the operation should proceed. The child died on the operating table the next day.

It was not until April 1995 when the scandal was leaked and came out on the front page of the *Daily Telegraph* that there was any action. Dr Bolsin's employment was threatened and he was directed to 'reduce contact' with the surgeons in question. The Department of Health's initial report (by another paediatric surgeon) stated that Dr Wisheart was a 'high risk surgeon' and should not carry out any further paediatric surgery. The CEO, however, referred to the report as 'a draft working document' and asked for alterations to be made. A second report was issued stating that 'concern by anaesthetists had raised anxiety among surgeons and this may have contributed to higher mortality'. So, Wisheart continued to operate on children. In May 1995, he undertook his last paediatric CT procedure – the child died a few weeks later.

In February 1996, Dr Bolsin moved to Geelong in Australia – as a whistleblower, it was too difficult for him to work in the UK. In April 1996, he wrote to the General Medical Council and requested a proper inquiry into the events at Bristol, and suggested that the behaviour of certain doctors amounted to serious professional misconduct. Dr Bolsin was the only one to do this. An inquiry commenced in 1996 and concluded in June 1998. The CEO (Dr John Roylance) and the two surgeons (Dr Wisheart and Dr Dhasmana) were found guilty of serious professional misconduct. Roylance and Wiseharst were struck off the medical register. Dhasmana was forbidden from paediatric surgery for three years, but allowed to continue operating on adults.

This incident triggered the global Clinical Governance movement for patient safety and caused ripples in healthcare everywhere. It highlights failures of individuals to act, system failures and failures to act along the entire 'chain of command', including the Department of Health, and right up to the President of the Royal College of Surgeons. It also illustrates the tendency of the medical hierarchy to close ranks and protect their own, no matter the consequences. Many of us who know the healthcare system do not believe much has changed since the events at Bristol.

Under good leadership, most doctors will behave ethically. Under bad leadership, most will not. Just look at the contrast between the Nazi doctors and the Dutch doctors who stood up for what was right. The medical community is very good at running disinformation and counter-narratives when it suits a vested interest, because we know that people trust doctors. When large gangs of doctors become empowered and enabled by political favour, they grow in numbers very rapidly and their influence can be substantial. And those who harm human beings through collusion, inaction or worse, directly harmful actions, will aggressively shoot down the truth, attack truth tellers, and collude in obfuscation of data that would otherwise expose their lies.

10

EPIDEMIC DETECTIVES

IN 2000, AN INTERESTING OUTBREAK OF THE BAC-
terial infection tularaemia occurred in Martha's Vineyard,
Massachusetts. Fifteen people presented with the infection,
11 of them with the pneumonic form. Since the pneumonic
form is rare in natural tularaemia, it immediately raised
suspicion of a deliberate release. Prior to that outbreak,
pneumonic cases had only been seen once before in the US,
in 1978. In the earlier outbreak seven people who were in
the same cottage became infected. The virus was thought to
have been transmitted by a dog which aerosolised the bacteria
by shaking its wet fur. The bacteria in question, *Francisella
tularensis*, is found in soil, wild animals, ticks and sometimes
in water. Following the 2000 outbreak, a classic epidemic
investigation was undertaken, in which infected cases and
uninfected people in the same location were compared and
surveyed for various risk factors. It turned out that lawn
mowing was a risk factor, so in this particular case the
Martha's Vineyard epidemic was thought to have been spread
by cut grass contaminated with the bacteria being carried
through the air. Who would have thought it?

It was only through epidemiological analysis and investigation that a likely cause for the outbreak was determined. The investigators from local and federal health agencies published the investigation in the *New England Journal of Medicine*, which is an instructive case of why traditional epidemic analysis is important. Sadly, today we see much postulating and pondering about phylogenetics as the only evidence of origins of pathogens, but in truth, with gain-of-function research, the phylogenetic tree cannot often readily tell you about origin.

Origins of epidemiology

The first epidemic detective was a physician by the name of John Snow. In 1854, during an outbreak of cholera in London's Soho, Snow mapped cases, and realised that the disease was being spread by a water pump. Until that point, the theory had held that cholera was spread through 'miasma' in the air. When the handle of the water pump was removed, the epidemic stopped. Snow is today seen as the father of epidemiology, a science that began around epidemics but is now widespread across all areas of health, including non-communicable diseases.

Epidemiology engages various methods, among them randomised clinical trials, used for every approved new drug or vaccine today, and observational epidemiology, which includes case-control studies and cohort studies. The link between smoking and lung cancer, for example, was first uncovered by Austin Bradford Hill and Richard Doll in the UK using a case control study published in 1950. This was followed in 1954 by a cohort study that followed up British doctors at regular intervals for 50 years, and confirmed the link between lung cancer and smoking after 10 years of follow-up in 1964.

The field of epidemiology has expanded substantially since then, and underpins all of medicine and health today. Bradford Hill also established the Bradford Hill criteria for causation, which I was taught as a field epidemiology trainee in the early 1990s, but which many experts are ignorant of. There has been much pontification by experts about 'mystery' diseases, including enterovirus D68 which has caused a polio-like syndrome in multiple countries since 2014. While this pontification was ongoing, we published a paper using the Bradford Hill criteria to show it was likely that EV-D68 was the cause of polio-like paralysis. I believe these were the first signs that since the 2009 influenza pandemic, the hijacking of public health by people without the requisite knowledge and training was well underway. The 2009 influenza pandemic showed that pandemics are big business, providing many funding and other opportunities which had dried up for other areas of infectious diseases such as HIV. The ensuing years saw this gradual shift in composition of expert committees and decision-making in many countries, which was exposed 11 years later during the COVID-19 pandemic. So, whilst epidemiology as a discipline expanded to underpin all of medicine and health, the specialised skills of field epidemiology remain in short supply, with no recognition of how vital this is in a pandemic.

Epidemic investigation – the science of investigating and outbreaks – was pioneered in the United States with an elite training program, the Epidemic Intelligence Service in 1951.

This program was the brainchild of public health physician and chief epidemiologist Alexander Langmuir. The Communicable Disease Center, founded by Langmuir in Atlanta in 1946, itself grew out of a US military program to control malaria during the Second World War. At that time malaria was mostly endemic in the southern parts of the US. In 1951 malaria was eliminated from the US – the same year the Epidemic Intelligence Service was formed. Today the Communicable Disease Center is known as the Centers for Disease Control and Prevention, and has evolved and grown to be the largest public health organisation in the world. Established at the start of the Korean War, the initial remit of the Epidemic Intelligence Service was to detect biological warfare. Langmuir pioneered the science of field epidemiology, also called shoe leather epidemiology, which sees trainees engaged in short bursts of technical learning in the classroom, and extended periods of time in the field investigating outbreaks and applying that learning to real problems.

The Epidemic Intelligence Service spawned a global network of Field Epidemiology Training Programs, including one at the Australian National University (ANU), which I undertook in 1992–93. It was the most ground-breaking form of learning I have ever experienced, and shaped my entire career. Through this training, I learnt the science of epidemic investigation and control, including what data to collect in an outbreak, understanding the epidemic curve, case finding through testing or using a clinical case definition, contact tracing, and doing a case control study to determine the cause of an outbreak. I did intensive coursework blocks on campus, where we were put up with our classmates in lovely townhouses overlooking Lake Burley Griffin and the Molonglo River. Bonding with the cohort was part of the experience, and when we finished our intensives, we would disperse around the country to our field placements. I completed mine in Melbourne, working at the Victorian Department of Human Services. This was a fantastic experience, where I did numerous outbreak investigations, often in little country towns. We would usually go in pairs in a government car to investigate outbreaks, speaking with locals and health services to understand the situation and gather data. In one outbreak in the Victorian town of Bright, we inspected the town water supply and found an animal carcass in the water, the probable cause of the outbreak. In 1992 I did my first aged-care facility outbreak investigation, the start of a long research career into such outbreaks.

Sometimes we had to set up surveillance systems to monitor cases and track progression of an outbreak. When in 1993 unprecedented floods occurred in Victoria in the towns of Benalla and Echuca, and surrounding areas, my fellow trainee Mark Veitch (now director of public health in Tasmania) and I drove up there from Melbourne to set up a surveillance system for flood-related outbreaks and other illnesses, both immediate and delayed. The immediate effects include injuries, shortages of medicines, hypothermia, carbon monoxide poisoning from use of gas generators and worsening of chronic diseases from loss of medication or lack of access to healthcare. The delayed effects include mosquito-borne infections, gastro, mental health problems, and worsening of epidemic infections like influenza or COVID-19.

Public health surveillance

Public health surveillance is key to controlling disease. If COVID-19 had been detected early, when there were only a handful of cases and before it had spread outside of China, the pandemic could have been prevented by isolating cases and quarantining their contacts. A surveillance system monitors trends in disease, so that you can respond to an increase or know by a decrease that control measures are working. Traditional public health surveillance relies on laboratories and doctors to report notifiable infectious diseases. Australia has over 70 nationally notifiable infectious diseases, and there is a statutory obligation for labs and doctors to report. Formal surveillance is often delayed, because it goes through validation and checking. Epidemic diseases can spread very rapidly, so there is value in detecting them very early. Syndromic surveillance systems, which provide immediate signals and feedback on existing health data in real time, can detect epidemics earlier than formal surveillance. These systems rely on doctors in emergency departments or primary care reporting on a specific syndrome, such as pneumonia or influenza-like illness. Even earlier signals can be detected using open-source data. When outbreaks occur, people talk about them on social media or news agencies report them long before health authorities know about them. In countries where health systems are weak, or information is censored, rapid epidemic intelligence can spot outbreaks early and overcome poor reporting or censorship.

During the COVID-19 pandemic some obfuscation of data has occurred in Australia, especially around deaths and the suggestion people died with the disease rather than from it, and that deaths occurred in less 'valuable' people. When the COVID Omicron wave surged in late 2021 in Australia, testing capacity was exceeded. In response, eligibility for PCR tests was restricted. The other alternative, rapid antigen tests, are unaffordable to many. As a result, testing rates are estimated to be low and reported infections are a likely underestimate. The dismantling of COVID-testing infrastructure in Australia in 2022, means we are ill-equipped to detect serious new variants of concern or to get early warning of a surge that may stress the health system. Through March and April 2022, COVID-19 hospitalisations in Australia rose steadily, but case numbers did not show a clear trend, sometimes even appearing to decrease. This is because many cases are not being captured due to the low testing rates. In a situation with poor case surveillance, only the hospitalisation data are informative of whether the epidemic is growing or falling. The WHO advises that the positivity rate of tests of about 5 per cent or less indicates a good surveillance system for COVID-19, and anything higher indicates inadequate testing. In April 2022, the test positive rate for Australia was 21 per cent.

More advanced surveillance with risk analysis methods enables us to differentiate different patterns of disease. In the era of engineered viruses, how do we recognise biowarfare and differentiate natural emergence of new strains versus man-made modifications to pathogens? To determine whether an epidemic has a natural or unnatural origin, firstly, the question simply has to be asked. It is basic logic that if we never pose the question as to whether an epidemic is unnatural or natural, we will never identify unnatural ones. This is obvious and yet it constitutes probably the greatest barrier for health professionals, because it is not something we are taught or trained to consider. We are taught to respond, treat and prevent epidemics, but never to question if they are unnatural. We are also taught to ridicule any suggestion of unnatural origin as a conspiracy theory. We assume everything is natural, and when someone suggests an unnatural origin, we, as a profession, tend to get very hot under the collar and loudly shout down that person. We saw that in the Rajneesh attack, during Sverdlovsk, during the Russian influenza epidemic and similarly when lab leak theories were raised about SARS-CoV-2.

What does a biological attack actually look like? Most people imagine it resembling a zombie apocalypse, with people dressed in hazmat suits and sirens blaring. Yet in truth, a biological attack does not come with a neon sign saying 'Unnatural' or 'Warning! This is an attack'. A biological attack, unless it is smallpox, which has been eradicated, will most likely pass for a natural epidemic. A syndrome is the clinical presentation of illness. There are only a handful of clinical syndromes caused by serious infectious diseases – and they are familiar syndromes like severe pneumonia, rash and fever, gastrointestinal syndromes and neurological syndromes like meningitis, encephalitis or paralysis. Pneumonia is a syndrome, but may have many causes including influenza, COVID-19, pneumococcus, legionella or mycoplasma. Paralysis may be caused by a road traffic accident, a stroke, by polio or Enterovirus D68. Were a biological attack to take place, we would see people presenting to hospital with one of these familiar syndromes, and because the number of cases would be higher than the baseline levels normally seen, health authorities would eventually realise it was an outbreak. However, if the outbreak was of a new disease, the normal tests conducted for that syndrome would come back negative and the disease might present unusual features, as was the case with COVID-19.

An epidemic is an outbreak at a larger scale, and a pandemic is an epidemic that spreads globally (when people say 'global pandemic' it is a tautology). Some infectious agents that would be used in an attack, like anthrax, do not spread from person to person, which means the impact is limited to the area of attack and does not keep growing. In the US anthrax attack of 2001, even though the case numbers were small, several were inhalational anthrax, which generally indicates an unnatural exposure, and the source was quickly pin-pointed to letters. This resulted in closure of the United States Postal Service and a massive decontamination operation in multiple buildings. Infectious diseases that do spread from person to person are more difficult to pick as unnatural, because, if they are highly contagious, once the spread starts, a natural and unnatural outbreak would look the same. The origin can only be determined by a forensic, intelligence and law enforcement investigation.

Early warnings

EPIWATCH, the AI-based epidemic observatory I developed since 2016, uses open-source data such as news feeds and social media to generate early warning signals for potentially serious epidemics. The event that inspired this was the 2014 West African Ebola epidemic. Through extensive research, testing and evaluation of the system, we have shown that

EPIWATCH is able to detect valid early warning signals for epidemics. EPIWATCH searches in over 40 languages, including every major Asian language. It provides a free public dashboard with a searchable, sortable world map and 30 days of data. In 2021, a grant enabled us to scale up EPIWATCH, and in 2022 a donation from Balvi, a philanthropic fund established by co-founder of the cryptocurrency Ethereum, Vitalik Buterin (whose vision is to decentralise money, and enable open access and open sourcing), allowed us to start to make the system globally accessible and open sourced. We will also be developing dashboards in other languages – in Asia in the first instance – so that it can be used at the grass-roots level in local communities. EPIWATCH's goal for public health is to prevent serious epidemics and pandemics. In order for it to have any prospect of preventing the next pandemic, the tool needs to be widely used around the world. This means it cannot be a gated, elitist or expensive tool, but needs to be made available widely, so it can be used at local levels in countries everywhere.

Data can be censored and manipulated by governments to suit political agendas. We have seen this occur during COVID-19, with some governments ceasing to report on case numbers, or not making testing accessible, with the effect that officially reported case numbers appear lower than they actually are. In Australia, there was no formal reporting of health worker infections with COVID-19, and substantial effort was put into minimising the impact on health workers, including opaque attribution of source of infection to 'community transmission'. For her PhD, one of my PhD students, Ashley Quigley, conducted analysis of open-source data. Collecting news reports detailing health worker infections and hospital outbreaks, she then analysed these to create a comprehensive picture of the impact on health workers and the health system in 2020 – something that was never provided by government. We use this same philosophy in EPIWATCH: we capture valuable intelligence which may not otherwise be reported. But with EPIWATCH, we do it continuously and in real time.

Despite the valuable potential of AI technology, many countries have been slow, even reluctant to adopt it in public health. Epidemics grow exponentially, over days or weeks, therefore rapid epidemic intelligence can make a vital difference to the speed and effectiveness of the response. The space of a few days or a week can make the difference between an epidemic being stamped out or spreading beyond reach. In several research papers, we have shown that EPIWATCH can detect epidemics early. In one paper published with Dr Cecile Paris and her team at CSIRO's Data61 and their post-doctoral student Adi Joshi, we showed the 2014 Ebola epidemic could be detected months before the WHO knew about it.

In further research with Dr Paris and her team, we showed we could have got a signal for the fatal thunderstorm asthma event in Melbourne in 2016 nine hours before any health officials were aware of it. During that thunderstorm, ten people died, and over 3400 people were hospitalised in a single night. Unlike an epidemic, where a virus has an incubation period and the disease may take days or weeks to develop, this was an immediate event, where the storm caused inhalation of substances which were instantly hazardous for human lungs, in this case thought to be pollens broken up by the storm into much smaller fragments than normally present in the air. A nine-hour heads-up would have been enough for health authorities to issue immediate alerts that asthmatics should have their inhalers on hand and avoid outdoor exposure.

When I teach the course on bioterrorism and health intelligence, I go through the red flags for an attack. You may see precursor attacks of smaller scale preceding the main attack. This is because the method of release or even the agent of choice may not be certain yet, and small-scale attacks that would likely pass as natural are a way to do beta testing. Beta testing is a way of testing and identifying problems with a product before the final launch. You may even see 'testing' of naturally occurring agents on the Category A list such as plague, anthrax and tularaemia. If genetic engineering is involved, you may see multiple outbreaks of related but different pathogens ('field testing'). If the perpetrator desires to inflict a particular syndrome such as pneumonia, then different pathogens that cause pneumonia may be field tested to see which has the greatest impact. In this case you may see multiple outbreaks of a desired syndrome (for example, pneumonia, encephalitis) caused by different pathogens, clustered in time. In the case of a new or engineered pathogen, you may see multiple genetic variants detected at same or similar time, even in the same outbreak. One finding that should raise suspicions is the appearance of many dead animals or birds in a particular location. Initial testing of pathogens might be carried out on animals, and their bodies dumped. In 2015 a news report told of 40 sick cats and seven sick dogs having been dumped in bushland on Queensland's Sunshine Coast. While it was considered a newsworthy event, it is not the sort of intelligence that any law enforcement or intelligence agencies collect or even think about.

Finally, to detect unnatural epidemics, we can use the science of epidemiology. Initially, a traditional epidemic investigation should be done. This involves going to the location, interviewing people, developing what is called a 'line list' of all the sick people and their characteristics, and collecting the same data on healthy people in the same geographic area. We then look at probable exposures or risk factors, based on what the syndrome is, and get lab testing done to make a diagnosis. In the case of an outbreak of gastroenteritis, we would ask for a food history. For an outbreak at a restaurant or a private event, we would find out what was served, and ask everyone which of the available foods and beverages they themselves consumed. Through comparing the sick and well people we can calculate the odds of a person getting sick based on each type of food or drink they've consumed. We look for the size of the association and whether it is statistically significant. We also plot the epidemic curve, which is the number of cases per day over time.

The shape of the curve can be very informative. A one-hit event (referred to as a point source outbreak) – such as, for example, a food-borne outbreak from a contaminated food being served at an event, or an anthrax attack – will typically produce a bell-shaped curve: it goes up, peaks and then goes down. A contagious pathogen that spreads person to person, on the other hand, will form a propagation pattern with a peak and then oscillations with recurrent peaks – much like we see with COVID-19. In a straightforward outbreak, we collect data from cases (that is, people who are ill) and controls (people who are not ill but are otherwise similar to the cases), and then do a case-control analysis to identify the risk factors for illness. Natural outbreaks have predictable behaviours, so one way of analysing the origin of outbreaks is to look at the epidemic pattern. In an unnatural outbreak, a discrepant epidemiologic pattern – or unexpected behaviour of the pathogen given its known characteristics – can be observed. Sometimes just a standard outbreak investigation is sufficient to sort out the cause, even if it looks unusual to begin with, like the tularaemia epidemic in Martha's Vineyard.

The mathematical term R0 is the reproductive number, and a key parameter for responding to outbreaks. It is the number of secondary infections caused by a single case in a population with no immunity. For example, if a person infected with measles walks into a room of 100 people who have never had measles or been vaccinated against the virus, and 20 of them catch the infection through the exposure, the R0 of measles is 20. Rt, on the other hand, refers to the R value when immunity or public health interventions are present to reduce R0. So, if half the people in the room are immune to measles due to vaccination or past infection, the infected person would only infect ten people. The Rt then in this case is 10. If a lockdown meant fewer people were mixing in society, or if mask mandates were being used, the Rt would also drop. All public health interventions we use to control epidemics (including vaccination, masks, testing and tracing, density limits, and lockdowns) aim to reduce the R value below one, and thereby stop the epidemic. An epidemic infection is one which has an R0 greater than 1. This means that it has the innate ability to grow exponentially over days or weeks and cause an epidemic. If R is less than 1, however, the number of cases decreases, meaning that infection cannot be sustained and dies out. Therefore R=1 is called 'the epidemic threshold'.

When we use public health measures to 'flatten the curve', this reduces the Rt and dampens the epidemic peak.

Risk analysis tools and forensic investigation

Proving a biological attack can only be done by law enforcement or intelligence agencies. This is because definitive proof requires the same forensic investigative capabilities as a homicide investigation. There needs to be evidence, such as finding a clandestine lab, or records from such a lab, identification of perpetrators, and uncovering a motive or accident. In the Rajneesh case, the motive was so unusual that only a full investigation would have uncovered it. Public health experts can only point to unusual patterns, but usually don't. Bioterrorism can rarely be proven by public health authorities alone – but epidemiology can help. Public health authorities have capability to analyse epidemic patterns and identify whether they can be explained by known facts about transmission of that pathogen. Ideally, the contribution of public health professionals is to flag aberrant patterns and refer the matter to law enforcement agencies for investigation. In the Rajneesh case, however, the public health authorities were not prepared to consider the possibility of bioterrorism.

Risk analysis tools are available for analysing and differentiating epidemics. But most public health personnel are not aware of them. The most commonly used risk analysis tool is called the Grunow-Finke tool, developed to assess a tularaemia outbreak in Kosovo by two scientists from the German Armed Forces Medical Academy, Roland Grunow and Ernst-Jürgen

Finke. Grunow and Finke judged it to be a natural outbreak, but demonstrated you can use a systematic and scientific risk analysis tool to evaluate the origin of outbreaks. However, it has low sensitivity, tending to incorrectly classify unnatural outbreaks as natural. We did research on improving the sensitivity of the tool. My PhD student Jessie Chen tested and re-tested it, modifying each criterion until the sensitivity was 100 per cent. We published our findings on the modified Grunow-Finke Tool (GFT), and have used it to analyses important epidemics.

The GFT has been the most widely used method in military medicine, but is largely unknown in general public health practice, and certainly not known to the virologists who argue on Twitter about the origins of SARS-CoV-2. Listening to them, you would think the definitive answer to the origin of viruses is in the genetic sequence. In some cases, there may be obvious signatures of genetic manipulation. But more likely than not, there won't. This is because the existence of gain-of-function research (serial passaging of a virus through an animal host until it acquires new and enhanced functions) means that an unnatural virus will appear natural.

Genetic patterns of viruses can be analysed by plotting them on an evolutionary tree, with the root being the original virus and the branches reflecting mutations and evolution of the virus. This is called a phylogenetic tree. When a new virus appears and the phylogenetic tree says that it has been lurking around for years or decades, that should raise a question mark over possible gain of function. While virologists tend to emphasise phylogenetics as the only source of truth about a virus, in fact epidemiology can also be informative.

At our EPIWATCH observatory, we watch Ebola outbreaks closely. Genetic analysis of the 2014 Ebola outbreak showed that rather than the outbreak virus starting and then spreading from Guinea in logical chains of transmission, there were multiple and repeated introductions of the virus everywhere. Ebola has an R0 (reproduction rate) of about 2, which means that one would expect to see its spread occurring in concentric circles (with some satellite epidemics due to people travelling beyond their local area). Usually epidemics spread according to what's called the 'stone in the pond' principle – you see expanding concentric circles reflecting the spread from the initial case to their close contacts, and then the contacts of the contacts, and so on. Occasionally a contact may travel and seed a new epidemic in an entirely new location. What we observed with the 2014 Ebola outbreak was that the concentric circle spread was much less than the numerous satellite epidemics. There are two different ways of explaining this: one is a large number of undetected chains of transmission (which masked the concentric circles of spread), and the other is an unnatural event with multiple releases.

We have a preconceived notion that an attack will be something dramatic and frightening, but it may not be. It may simply be diarrhoea and vomiting in a lot of people, like the Rajneesh attack. Historically, most attacks have not been recognised at the time, because the question has not been asked and the outbreak has been assumed to be natural. Even in Sverdlovsk, which involved inhalational anthrax (the biggest signpost you could get that it was unnatural), the American experts were ready to believe the Soviet propaganda that it was a natural outbreak. Identifying bio-attacks is not a standard public health competency. Yet it is important we develop competency in this, so that we can respond in the most effective way. As we saw with COVID, the response may be politicised, data ignored and the health of the public devalued.

11

THE FUSS ABOUT FACEMASKS

I LEARNT THAT SCIENCE CAN BE SUPPRESSED BY politics in 2006 when I began studying the efficacy of N95 respirators and surgical masks. With outbreaks of avian influenza in Asia, there was a flurry of pandemic preparedness activity. The then chief medical officer of Australia, Dr John Horvath, approached me about designing a trial of masks in the community, as there were no published clinical trials available. It was the first trial ever to be conducted in the world on masks, and the second to be published. From there, I decided it was important to research N95 respirators and surgical masks in health workers, who are at high risk of exposure to infections. A surgical mask is a device designed to prevent splash or spray of liquid (such as blood splatter) on the face. It is loose and has gaps around the edges. A N95 or P2 respirator on the other hand is designed to fit around the face and filter over 95 per cent of airborne particles. For an airborne virus, a N95 or P2 protects best. Our own research and others during the COVID-19 pandemic showed there is a gradient of protection from no mask to cloth mask to surgical mask and finally N95. But in 2006 there was no solid evidence, only the first principles based on the design of each product. Guidelines by the WHO and most countries did not even mention cloth masks, although they were worn commonly in Asia.

In many Asian cultures, mask-wearing is common and health workers use masks regularly, so a clinical trial in such a setting is viable. At that time we had regular visiting delegations from China to the Children's Hospital at Westmead in Sydney, where I worked at the National Centre for Immunisation Research and Surveillance. I went along each time and pitched my idea of doing a randomised controlled trial on N95s versus masks in health workers. In 2007 I finally met someone who was interested in the research: Dr Wang Quanyi from the Beijing Center for Disease Control and Prevention, who leads an amazing team of smart, hard-working epidemiologists and scientists. My team and his formed a friendship over the many years of collaboration on clinical trials of masks and N95s in China, which resulted in the publication of over 30 scientific papers together.

When I entered the field of mask research, little did I know that I was stepping into a minefield of ideology and opposition dominated by hospital infection prevention and control (IPC) experts. These experts are usually doctors and nurses who specialise in antibiotic resistance and protecting patients from wound infections and other hospital-acquired infections. IPC as a discipline is focused on patient safety. As such, health workers are often perceived as the enemy that infect hapless patients by forgetting to wash their hands. The occupational health and safety of health workers themselves is not the main focus for IPC experts. And yet it is precisely these experts into whose hands the safety of health workers is placed every time there is a dangerous epidemic or pandemic. Since SARS in 2003, much of the IPC community had rejected airborne transmission of SARS, influenza and SARS-CoV-2, and denied airborne protection to health workers, treating them with hostility and disregard their safety.

A letter to Obama

During the 2009 pandemic of H1N1 influenza, as one of the few researchers who had done a clinical trial of masks, I was invited onto a US Institute of Medicine (IOM) committee on personal protective equipment (PPE) for health workers. The guidelines we formulated recommended N95 respirators for health workers, and on communication of these recommendations I suddenly felt the full fury of the US IPC community. The three peak IPC organisations, the Infectious Diseases Society for America, the Society for Healthcare Epidemiology of America and the Association for Professionals in Infection Control and Epidemiology, all got together and wrote a letter of protest to President Barack Obama. Their letter, which specifically named me and blamed my research for the new guidelines, demanded the immediate reversal of the new Centers for Disease Control and Prevention guidelines. Outraged that N95 respirators would be recommended for health workers, they demanded that this be retracted and that less protective surgical masks be recommended instead. My research had produced an unpopular finding: N95s (which are designed as respiratory protection) protect against respiratory infections, while surgical masks (which are designed to stop splashes of fluid or to stop a surgeon contaminating a wound) do not offer good protection. My research had challenged their decades-old dogma, and so created outrage among the infection control community globally.

The letter to President Obama alleged that my study was directly responsible for the IOM committee recommendations. It went on to attack my study based on a different presentation of the same data at two conferences that year. In fact, the IOM recommendations were not based on my research (which had not yet been published at the time), and it was explicitly stated in the report that only published data informed the recommendations. The Centers for Disease Control and Prevention's peak body on worker safety, the National Institute for Occupational Safety and Health, wrote a counter-letter debunking the falsehoods in the letter. And so I found myself smack bang in the centre of a global controversy. I started receiving hate mail, mainly from people in the IPC gang: 'You are in the pay of 3M' (a leading manufacturer of N95s), one piece read. 'Your science is like the MMR-autism theory,' read another (ironic, as I have worked in vaccine research my whole career and published several papers specifically debunking the MMR-autism theory).

Another letter said: 'You had a conflict of interest being on the IOM committee.' In fact, as the only foreigner on that committee, with no vested interest in American healthcare, I did not have a conflict of interest. The comment that took the cake was: 'Do you know what you have done? Hospitals in the US are now subject to fines for not using N95 respirators during contact with patients infected with novel H1N1 influenza.' In actual fact, I had not done anything to this aggrieved soul. I had merely done a piece of research that shook the foundations of a long-held ideology about masks, and for my sins had been on a committee that drove PPE policy for American hospitals during the 2009 pandemic. Suffice to say, the angry IPC community overturned the Centers for Disease Control and Prevention guidelines, leaving health workers with only surgical masks as a protective barrier during the 2009 pandemic. And I myself became a thorn in the side of the IPC community forever more, including in Australia.

Our paper which demonstrated the superiority of N95s over surgical masks in health workers was in process for publication in a leading journal, placed alongside another, much smaller study that showed no difference between these devices. The journal reviewers even asked us to 'tone down any reference to N95s being superior' and to omit the data on the control arm, which we did. We also tried to downplay as much as possible the indisputable finding that N95s were superior. The identical data minus the control arm (aligned with the paper) was subsequently presented at a conference later that same year, and this was what the enraged IPC crowd jumped on. Meanwhile, the journal withdrew our paper, and only published the much smaller study which aligned with established IPC ideology. Those who opposed the IOM guidelines also actively promoted the results of that small trial. As a result, the smaller study dominated all discourse on health worker protection for over a decade, essentially all the way through to the COVID-19 pandemic. Our N95 trial, meanwhile, was blocked everywhere else we tried, but was eventually published in 2011. In the meantime I kept my head down and conducted several other clinical trials of masks and respirators, a substantial body of work that was ignored and minimised by the IPC community. The tide finally turned in 2020, when the stakes became too high and consequences too great to expect health workers to treat patients with a deadly disease in inadequate protection. In 2021, the WHO estimated that at least 180 000 health workers had died of COVID-19 in the line of duty.

Public Enemy No. 1

After 2009, I remained Public Enemy No. 1 to the IPC community despite keeping a low profile. They continued to attack my research, and to make it difficult for me to get subsequent clinical trials published. When in 2014, in the midst of the West African Ebola epidemic, the WHO IPC committee issued their 'evidence-based' guidelines for health worker PPE, they completely overlooked my research, despite the fact that three of my trials – by far the largest and most rigorous trials done to date – had by then been published. When I wrote to the WHO, querying why their mask guidelines omitted to cite my published clinical trials, I got a reply saying that they were busy with the Ebola response and would get back to me later. I replied to them, telling them that the Ebola crisis was not an excuse to postpone a response. That, on the contrary, it made a response even more urgent. I reminded the WHO that they had a responsibility to ensure optimal occupational health and safety of healthcare workers, and that evidence clearly indicating the greater efficacy of N95 respirators over surgical masks needed to be taken into account.

There were clear inconsistencies in the recommendations laid out by the WHO and other agencies. On the one hand, they recommended a N95 for lab workers handling Ebola, but, on the other, they recommended only a surgical mask for health workers treating Ebola patients. A lab is a highly controlled, sterile environment, whilst a hospital is unpredictable, dynamic and far more hazardous for workers – a patient can vomit or cough on you without notice, or pull out their drip and spurt blood on you. In 2014 that is exactly what happened with Patrick Sawyer, the index patient in the Nigerian Ebola epidemic. Sawyer, who travelled from Liberia while in the early stages of Ebola, was hospitalised in the First Consultants Hospital in Lagos, and wanted to be released to continue his travels. Resisting hospital staff, in one attempt to leave, he pulled out his IV, spurting blood on staff. The outbreak caused by Sawyer resulted in 18 infections and the deaths of eight hospital staff. Dr Ameyo Stella Adadevoh, who was responsible for preventing Sawyer from leaving the hospital, died of Ebola, despite having had no direct contact with him. Adadevoh is credited with having prevented a major epidemic in Nigeria.

We then wrote a paper challenging the WHO and US Centers for Disease Control and Prevention PPE guidelines for Ebola. We called for airborne precautions, but the top medical journals would not publish the paper. Instead they published pieces praising the guidelines of the Centers for Disease Control and Prevention and the WHO. One such piece said, 'Using extra gear inflates patients' and caregivers' anxiety levels, increases costs, and wastes valuable resources. More insidiously, requiring precautions that exceed the Centers for Disease Control and Prevention's recommendations fans a culture of mistrust and cynicism about our nation's public health agency.' That suggests the reputation of the Centers for Disease Control and Prevention is more important than the safety of health workers. Another such piece went as far as to say that 'In fact, goggles and masks might not even be necessary to speak with conscious (Ebola) patients, as long as a distance of 1–2 metres is maintained.'

I knew that the way forward was not going to be through the medical establishment. I got in touch with Professor Trish Davidson, whom I had known a long time and who is now the vice chancellor of the University of Wollongong. At the time, she was the dean of Johns Hopkins Nursing and very influential in the field. I also contacted an intensive care physician from South Africa, Dr Guy Richards, who has extensive experience with clinical management of haemorrhagic fever and with whom I had engaged in discussions about PPE for viral haemorrhagic fevers (VHF). Dr Richards, who works at the Charlotte Maxeke Johannesburg Academic Hospital affiliated with the University of the Witwatersrand, told me that when treating VHF they always used powered air purifying respirators (PAPRs) which are the Rolls-Royce of airborne precautions, and much more comfortable for health workers than a disposable N95. Then we contacted Professor Ian Norman from King's College London, a nursing academic leader and editor-in-chief of the *International Journal of Nursing Studies*. He was more favourable toward considering our paper. After undergoing peer review the paper was finally published in September 2014. It was the first piece to call out the WHO and Centers for Disease Control and Prevention on their inadequate Ebola PPE guidelines for health workers. To this day, the WHO continues to recommend a surgical mask for health workers treating Ebola. Strangely, their 1999 guidelines recommended a N95, but for unknown reasons they decided to downgrade the protection for health workers in 2014. The Centers for Disease Control and Prevention, however, were forced to upgrade their guidelines after two nurses became infected with Ebola in the US.

The battle for public acknowledgement that N95s are superior to surgical masks has been a long one. The IPC community has generated counter-narrative after counter-narrative, often in the form of biased systematic reviews, to maintain that surgical masks are equally effective. As the only researcher whose group had published large trials of N95s for most of the decade since 2009, I was standing alone against the whole global IPC community and the journals they commandeered. In Australia, my research into masks has put me in an invidious position among infectious disease physicians. The majority viewed me with a combination of resentment and fear. On one occasion in 2022, an Australian closely tied to the infectious diseases community whose wrath I had incurred, banned me from speaking at a conference he was organising. I had been invited by a European company to speak at a session they had sponsored. The sponsor reported back to me that the Australian organiser said, 'Raina MacIntyre cannot speak at this conference.' When the shocked sponsor asked why, the man replied that I had angered one of the main conference funders. I looked up the conference website and saw that the two main funders were the Australian government and a foreign government with whom I had no dealings. This hints at the politicisation of our pandemic response, with some experts 'in favour' and others blacklisted.

For over a decade I appear to have been blacklisted in Australia, generally not invited to speak at Australian infectious diseases conferences, whilst I am regularly invited to speak at much larger international conferences. The same with committees, including those on making guidelines on PPE, when no one in Australia comes close to my expertise and research track record on masks. Nor has other relevant expertise (such as Lidia Morawska) been utilised for the national benefit. Instead, committees are stacked with IPC experts who deny airborne transmission of respiratory viruses, refuse N95s to health workers and promote handwashing as the most important solution to respiratory virus transmission. When you think about it, this is as ridiculous as saying a gastrointestinal infection is best prevented by opening the windows. When monkeypox broke out, the denial that monkeypox could sometimes be transmitted by the respiratory route began in earnest. The WHO and Centers for Disease Control and Prevention quietly changed their long-standing advice that monkeypox is airborne to say it was not. In Australia, despite being the only one to have published research on monkeypox prior to 2022, and on a WHO expert working group on smallpox and monkeypox, I was not asked to assist with the Australian response.

Crunch time

The mask controversy of 2009 came to a head in 2020. By then it was no longer flu but a new virus, where the consequences of incorrect policy were much more likely to be fatal. Too much was at stake for the IPC experts to keep getting away with it, and too many people were directly affected and terrified for their own safety. During the pandemic many were denied adequate protection. One of my clinical trials was the only published study on cloth masks when the pandemic hit. After working in China on the first two trials, I realised many health workers used cloth masks. Yet cloth masks had been entirely ignored in policy documents, with the assumption that disposable masks would always be available. This became the focus of intense interest in 2020 when mask shortages forced health workers to wear home-made cloth masks to treat COVID-19 patients. My trial showed people using a cloth mask had worse outcomes than surgical masks and, surprisingly, worse than using no mask. A frantic doctor contacted me from the US, asking, 'Is it better I wear no mask to treat COVID patients?' Horrified, I wrote back and advised that he should not work without a N95, but that if he chose to work, a cloth mask was better than nothing. We then analysed additional data from the trial to find out why cloth masks performed poorly – and found it was due to inadequate washing. A cloth mask needs to be washed daily in a washing machine at high temperature, and if so, can be as good as a surgical mask.

Those healthcare workers who tried to bring their own respirators to work were harassed, bullied and refused permission to protect themselves better. Many healthcare workers contacted me in distress, saying the IPC team were policing them and pouncing on anyone daring to bring their own N95 respirator, while hospital supplies were under lock and key. When third and then fourth doses of COVID-19 vaccines were finally introduced in Australia, health workers were not prioritised. Imagine being on the frontline, being sickened by COVID in the line of duty, and yet being denied protection for the occupational risk you face. Time and again, the IPC committees denied health workers a N95 or P2 respirator for protection. One courageous infectious diseases physician, Dr Michelle Ananda-Rajah, advocated at great personal cost for fellow health workers' safety and, together with psychiatry registrar Dr Ben Veness, formed a group called Health Care Workers Australia. Deciding that the system was so broken she needed to change it from the inside, Ananda-Rajah ran for and won a seat in federal parliament in the 2022 election, despite concerted campaigns to smear her in the mainstream media. In her moving maiden speech to parliament, she advocated again for health workers.

Most countries only stockpile a few weeks' supply of medical items, but pandemics can go on for years. In July 2019, I published research on the health system's impact of a smallpox epidemic in Sydney. In that research I showed that for Sydney alone, we needed at least 30 million N95s or similar respirators stockpiled for such an event to provide sufficient cover for a period of at least a few months. I sent the paper and media release to the then chair of the Communicable Diseases Network of Australia (CDNA), a loose coalition of State and Territory health departments. She brushed it off and told me how well prepared Australia was. That same person also said that it was a shame I did not reference the smallpox SoNG (Series of National Guidelines) in my paper and remind the world of how excellent it is. In fact, the SoNG was not released until November 2019 – four months after my paper was published. The SoNG failed to recommend any respiratory protection for smallpox, the most highly airborne pathogen ever known. Only contact precautions – gloves and apron. This was corrected after I emailed colleagues in the Department of Health and sent them some references, including a paper we had published on the unique propensity of variola to spread through the air over long distances. In July 2019 when I alerted the CDNA for the need for more N95s in the stockpile, there were about 3 million respirators stockpiled nationally – far less than the required 30 million for Sydney alone. In December 2019 we then had catastrophic bushfires, and widespread smoke exposure. This ate up more than half the paltry stockpile, none of which was replenished when the COVID-19 pandemic reached our shores. Moreover, Australia had sent masks and PPE to China in January 2020 to help the situation in Wuhan. So when GPs were asked in early 2020 to be the frontline of the COVID-19 response, they faced severe shortages of necessary protective equipment.

Principles of PPE

The US faced its first Ebola scare in 1989 when a group of macaque monkeys shipped from the Philippines to Reston, Virginia, for medical research began to die in their cages. Afflicted with internal haemorrhaging, the monkeys succumbed one after another. The disease, identified as Ebola, appeared to have spread through air vents in the building, as monkeys held in one room managed to infect others in a separate room. In the end, over 450 monkeys were euthanised. One monkey even escaped, causing a panicked hunt. The creature was eventually found. A pathologist at the United States Army Medical Research Institute of Infectious Diseases, Dr Nancy Jaax, was accidentally exposed to the blood of an infected monkey despite being triple gloved at the time. Jaax had cut her hand in the kitchen earlier and noticed a hole in her glove on the cut hand after being exposed to the monkey blood. Despite the exposure, however, Jaax did not get infected. It later emerged that the particular strain of the virus, which became known as Ebola Reston virus, does not cause illness in humans. In contrast, the Zaire strain which caused the 2014 West African epidemic killing over 11 000 people, is deadly to humans. For the Centers for Disease Control and Prevention to advise health workers in 2014 that single gloves offered sufficient protection against the Zaire strain is still unbelievable. Even without a visible cut on one's hands of the sort Dr Jaax had, a person's skin usually has many micro-abrasions that cannot be seen easily. Although they are invisible, these present an easy pathway into the body for viruses.

An ordinary person may never need to deck themselves out in full PPE, but in the era of COVID-19, knowing at least how to use masks properly is important. For police, paramedics and other first responders, the risks of exposure are far higher. First responders facing an unknown hazard should assume the worst and protect themselves accordingly. All skin, hair and mucous membranes should be covered, and the PPE should be made of water-resistant materials so nothing can seep through to the skin. Double gloving is recommended when treating patients with Ebola. Imagine the following scenario: you have just treated a patient with Ebola, and you are removing your PPE. Bear in mind that Ebola has an extremely high fatality rate. If you are only wearing one pair of gloves, you remove one glove, and then how do you remove the other? Only by touching it with the bare hand that has just been un-gloved, right? If you are double gloved, however, you can safely remove the contaminated outer pair, sanitise the first pair and then apply clean gloves over them to remove any additional PPE. Double gloving is critical to reducing contamination risk. This is especially important for high fatality diseases like Ebola.

This issue became a hot topic in 2014, during the West African Ebola epidemic. The US Centers for Disease Control and Prevention issued Ebola PPE guidelines recommending single gloves, no head or foot coverings, and the wearing of a surgical mask. Megyn Kelly, a star presenter for Fox News, took the then director of the Centers for Disease Control and Prevention to task over the guidelines. Megyn Kelly was featured in the movie *Bombshell* for her role in the take-down of the head of Fox News, Roger Ailes, over sexual harassment. Kelly, a whip-smart lawyer, asked the Centers for Disease Control and Prevention director if he himself would choose to treat an Ebola patient wearing just one pair of gloves, and with no PPE for his head and feet. The director dug his heels in and said he would. Kelly then put up an image of the Centers for Disease Control and Prevention director in full hazmat gear, covered from head to toe and double-gloved while visiting an Ebola treatment centre in West Africa. The Centers for Disease Control and Prevention changed their risky guidelines a month later after two nurses in the US became infected with Ebola while using their recommended PPE.

PPE worn for hazardous exposures includes gowns, suits, apron, boots or rubber shoes with shoe or boot covers extending up to the knee, head and neck cover, goggles or face shields, and surgical scrubs which can be discarded after use thereby removing the risk of contaminated clothes being taken home and exposing others in one's household. The putting on of PPE is called 'donning' and removal is called 'doffing'. The order of donning and doffing matters, along with the particular manner with which you wear the PPE, which can determine whether you contaminate yourself or not. Seemingly small issues like whether your first pair of gloves is worn under the cuff of your gown or on top of it also matter. It's essential to have a dedicated space for doffing and a protocol for reducing the risk of self-contamination. Attending appropriately to the waste generated from PPE is also critical.

Managing biological waste can be a major problem. It was faced at scale for the first time during the 2014 Ebola epidemic. In West Africa, where burial is the norm, directives ordering cremation and prohibiting burial rituals such as washing the body were culturally unacceptable. When the first cases of Ebola were treated in the US that year, the amount of biohazardous waste per patient was greater than any available receptacles. At Emory Hospital in Atlanta, which treated Americans who had been evacuated from West Africa, they had to buy 32-gallon drums to contain the waste – soiled bed sheets, clothes and used PPE. To make things worse, waste disposal companies refused to transport the hospital's waste. In some states, incineration (the best way to get rid of Ebola waste) is illegal. The logistics of handling massive quantities of biohazardous waste caught countries off guard. This was because their pandemic plans were based on influenza, where waste is not a consideration.

No beards please

As a defence against the spread of infection, the fit of a mask around the face is critical, because unfiltered air will flow through the gaps. The idea is to force air through the face piece, which is a filter. A beard precludes good fit because the mask cannot create a seal around the face. Bearded experts and health workers I know shaved their beards during the COVID pandemic. Professor Don Milton, for instance, whose seminal research on airborne transmission of pathogens such as influenza and smallpox I had followed long before the COVID-19 pandemic, always had a beard, but shaved it off during COVID-19 so he could wear an N95 mask correctly. In 2019, Milton accepted my invitation to attend the Pacific Eclipse international workshop, presenting his work on airborne transmission of smallpox. Later in May 2021 he tweeted: 'Raina MacIntyre – her Pacific Eclipse international tabletop exercise in Dec 2019 now seems almost clairvoyant. Her extensive research on the importance of respirators for HCW [healthcare workers] over decades, so often discounted and overlooked, is a signpost showing the way.'

In my professional career, I have engaged with military, police, other first responders and the intelligence sector. I have also published work on protection of police and other first responders. In early 2019, one year before the COVID-19 pandemic, I was invited to brief the police commissioners of

Australia and New Zealand on biosecurity. I have regularly been invited to speak at intelligence and national security conferences in Australia and the US, and in 2018 delivered a keynote address at the Five Eyes LinCT Counter Terrorism conference on new frontiers in biological threats. Five Eyes is an intelligence alliance between the US, UK, Canada, Australia and New Zealand. That same year I was fortunate enough to attend a chemical–biological threats training event held by the Australia-New Zealand Counter-Terrorism Committee held jointly by NSW Police, Ambulance and Fire and Rescue services. There, I was able to look at the kinds of PPE used and the standard decontamination methods available in the event of a hazardous exposure such as a bioterrorism incident. Specialist hazmat units are well prepared, but what about regular police?

Paramedics and police are often first on the scene and face unknown hazards before a diagnosis has been made. Police and paramedics also have a rescue culture – rushing in to save people is the nature of the job, but this also places them at high risk of hazardous exposures. Given the proliferation of clandestine labs, these days a raid on premises suspected of containing weapons may also contain drugs and biologicals. There are several cases of police exposure to fentanyl. In the Salisbury attack (see pages 202–203), one of the police officers, Nick Bailey was exposed to Novichok sprayed on a door handle. Despite wearing forensic PPE, he required intensive care and eventually left the police force. Law enforcement agents must assume the potential presence of biological substances in any crime scene they attend, yet in my assessment are not well prepared.

Many police were deployed to protests during the COVID-19 pandemic, wearing poor quality masks or ones that were incorrectly fitted. Often police are up close and personal with angry protestors screaming into their faces. Shouting and screaming generates vastly greater quantities of aerosols than talking or breathing. As such, this increases the risk of SARS-CoV-2 transmission. A N95 provided the best protection against this SARS-CoV-2 and this infection is worth preventing. The evidence is accumulating that reinfection increases the risk of complications, and may result in heart failure, lung impairment, dementia and other effects on the brain.

12

BRAIN EATING VIRUSES

THE IDEA OF A BRAIN-EATING VIRUS THAT ALTERS your brain function and changes who you are is a terrifying proposition. The zombie apocalypse is a popular theme in horror movies like *World War Z*. In the TV series *The Walking Dead*, a virus is responsible for turning people into zombies. In one haunting scene, zombies descend on the abandoned US Centers for Disease Control and Prevention in Atlanta. The common theme in these zombie apocalypse narratives is of an infectious agent that turns people into zombies. All it takes is a bite or scratch from a zombie for the affliction to be transmitted from one unfortunate person to others. In reality, many viruses do affect the brain, most commonly causing encephalitis, or inflammation of the brain. Meningitis, on the other hand, is inflammation of the outer lining of the brain, the meninges. Most infectious encephalitis is caused by viruses, for which there are no treatments. Meningitis, however, is usually (but not always) caused by bacteria, which can be treated with antibiotics. Encephalitis is infection of the cortical matter of the brain. Usually, infection has only temporary effects on the brain, but sometimes it may be permanent. Rabies is the most well-known virus that causes a fatal encephalitis 20–90 days after a bite from an infected animal. In cases of rabies infection, dogs or people foam at the mouth; this is because the virus affects the muscles involved in swallowing, resulting in inability to swallow. In addition, rabies causes agitation, anger, hyperactivity, hallucinations, confusion and muscle spasms. Rabies may be what has inspired the depiction of zombies in movies, except that it always results in death.

Measles causes encephalitis in one in a thousand infected people, and can produce seizures, agitation, confusion, behavioural and speech disturbances, and paralysis. In some cases, survivors are left with permanent disability. A rare and universally fatal complication of measles is a type of encephalitis called subacute sclerosing panencephalitis that occurs in one in 25 000 survivors many years after the infection. The illness may start with lethargy and irritability, but progresses to seizures and dementia, rigidity, movement disorder, mutism and eventually coma and death. Zika virus, too, can cause encephalitis in some people, but is better known for causing congenital abnormalities of the brain in newborns, specifically microcephaly (or a very small brain and skull), and permanent disability as a result.

Not just a cold

While the narrative that COVID-19 is just a cold or flu has been widespread, lulling people into a false sense of the pandemic being over, the evidence suggests otherwise. Early in the pandemic it was clear that SARS-CoV-2 can cause encephalitis and a range of other abnormalities of the neurological system, including loss of smell (due to damage to the nerves to the nose), strokes, and Guillain-Barre Syndrome, which causes paralysis. The SARS-CoV-2 virus causes massive vascular pathology, which results in blood clots and blockage of arteries, hence the risk of strokes and also heart attacks. One study published in the journal *Nature Medicine* showed that at least 12 months after infection the risk of heart attacks, strokes, clots and fatal abnormal heart rhythms increased by a factor of two. When a former Australian sports star tragically died in 2022, months after having contracted COVID-19, I wondered if he had suffered a COVID-19-related cardiovascular event. There was no discussion of this as a possible explanation, however, in public commentary.

The causes of COVID-19 neurological syndromes vary. But autopsy studies of people who died of other causes and did not have severe COVID-19 show that the virus can be found in the brain long after the initial infection. During the initial infection, some people develop strokes, encephalitis, meningoencephalitis, delirium or seizures. A substantial proportion develop 'long COVID' after the initial infection. Some of this is likely due to persistent, potentially irreversible effects on the neurological system. Studies of CT scans showed that after COVID the brain undergoes shrinkage. One study showed a drop in IQ equivalent to ageing 20 years, on average, in survivors of severe COVID-19. A brain once shrunken cannot grow back again, so these may reflect permanent changes. Another study showed pathological effects in the brain similar to those of Alzheimer's disease, while others show loss of myelin, the protective sheath around nerves in the spinal cord and brain. The 'brain fog' and cognitive difficulty reported in a large proportion of survivors of COVID-19 may well be a symptom of underlying brain pathology, and we do not know whether this will be permanent.

While the 'COVID-is-just-a-cold' narrative is being drilled into the minds of the people, Alzheimer's expert groups are already studying COVID-19's long-term effects and bracing for a huge increase in dementia secondary to the disease. Early studies show an increased risk of new-onset dementia after COVID. We do not know what long-term effect COVID will have on children and young adults with developing brains. Will we see a generation of people with diminished cognition as a result of the adults of today thinking that mass infection caused no harm? Abundant evidence of the effect of SARS-CoV-2 on the brain exists now to warrant a precautionary approach and avoid reckless mass infection, especially in children, who have the longest lifespan ahead of them and the most to lose. Tellingly, the same 'experts' who insisted that children had to go back to school before they were fully vaccinated – that even a few weeks' delay was unacceptable after two long years of home schooling – have been silent about potential long-term cognitive damage in kids.

'Long COVID' is a heterogeneous condition where symptoms persist after the acute infection. Studies vary in estimates of how common it is – anywhere from 4 per cent to 50 per cent – probably because the definition of long COVID varies between studies, and the syndrome itself is hetero-geneous. This is likely to be multifactorial: fatigue may be the result of brain effects, immunological effects, or subclinical heart or respiratory failure, and all or combinations of these may occur in survivors. Some of the neurological complications are alarming and have left sufferers in a diagnostic limbo. In addition to encephalitis during acute infection, a small proportion of people develop a frightening kind of brain inflammation called limbic encephalitis. Sometimes this is reported as COVID-19 psychosis.

A young doctor, Kelly Ann Fearnley, who graduated from medical school in the UK and started working as a doctor in early 2020, became infected in late 2020 while working on the COVID-19 ward. In an article published in *Medical News Today*, Dr Fearnley described a mild illness like the flu that lasted about two weeks. After four weeks, however, she had a range of persisting symptoms. She said she 'began having cyclic attacks of pins and needles in all four limbs and violent entire body shaking. The shaking was as violent as seizure, but it wasn't one because I was conscious.' She went on to describe how 'it was as if my body had forgotten how to breathe. It was at this time that I suspected something central was amiss, and recognized that I was poorly, even if no one else did. I do not doubt brainstem involvement.' Her symptoms were not diagnosed by the doctors she saw and were dismissed as anxiety. Then they became worse, and she started hallucinating:

> My sleep became disturbed. I felt a mental, physical,
> and emotional exhaustion like no other, but the virus
> or my body's response to it kept me wide awake. After
> 72 hours without sleep, I experienced external auditory
> hallucinations secondary to insomnia. Getting a glass of
> water I 'heard' a tannoy call me to my platform, whilst
> a man 'shouted' angrily at me from the corner of the

room. (A month later) I began having internal auditory hallucinations. I hallucinated 5 of 7 nights a week for 2–3 months. I would hear music, bands playing, news reports, political broadcasts, all originating from inside the ear. Although my sleep continued to be disturbed, these hallucinations were not secondary to sleep deprivation. When I was fortunate enough to sleep, I experienced night terrors; vivid, disturbing dreams.

The later hallucinations were due to encephalitis. After being misdiagnosed as 'anxiety' by a range of doctors, a neurologist eventually diagnosed Dr Fearnley's condition as limbic encephalitis, which is now a recognised complication of COVID-19. A known neurological complication of COVID-19 is loss of smell. The olfactory nerves in the nasal cavity are connected with the limbic system deep in the brain, which controls memory and emotions. Think of the brain as a giant cauliflower head: the outer parts are the frontal, temporal and parietal lobes, with the limbic system sitting deep inside those, at the 'core' of the brain and near the brain stem. It is not certain if the virus invades the olfactory nerves and continues on its destructive path right into the core of the brain, the limbic system, or alternatively, if it inflicts damage through immunological reactions.

The price of mass infection

During my time as a trainee physician at Royal Prince Alfred Hospital I did a number of neurology and cardiology terms.

These areas remained of interest even after I began doing infectious disease research in the early 1990s. During the COVID-19 pandemic, I began following the research studies that came one after another on both cardiac and neurological complications of COVID-19. Early in the pandemic, to help me keep track of the latest research and as a resource for others, I collected these papers on a COVID resources webpage. Right from the start, it was clear that COVID-19 was damaging to the heart, directly killing heart muscle cells, causing blood clots, vascular damage and a range of other complications. By 2021 it was apparent that COVID-19 also had serious neurological complications.

Of every infection I have covered, COVID-19 has the most concerning potential on the human brain and whole populations because of the sheer scale of infections. As I write, China is fighting off an Omicron epidemic, and I do not know if they will be successful. They are now among a very small number of countries that have kept COVID-19 under control; these include Taiwan and some small Pacific island nations that closed their borders. Geopolitically, taking a longer view of COVID-19, China may end up in a stronger position than the US over the long term if they manage to keep COVID out or limit its impact. We have not even begun to appreciate the population-level burden of chronic disease, cognitive impairment and organ damage that countries with a large burden of disease will be living with into the future.

Vaccines reduce the risk of severe complications, including long COVID, but alone are not enough to control the spread of SARS-CoV-2. Better management of COVID does not have to entail a choice between vaccination only (which countries like the UK and Sweden have chosen) and lockdown. All countries can use layered strategies to improve control of COVID, including vaccines, ventilation, masks, testing and tracing. But many have simply fallen into an all-or-nothing view. Protecting the population will have long-term benefits both for health and the economy. Studies have shown that during the pandemic the countries with the best health outcomes have also had the best economic outcomes. China has made a substantial investment in its public health system, hospitals and other health facilities. New outbreaks are tackled aggressively and, while much of the world has abandoned testing and tracing, in China there are high rates of implementation of these measures. Despite warnings from the WHO to strengthen testing and tracing, one after another high-income countries have dismantled their infrastructure. China has also continued using lockdowns, but if community transmission becomes established in China, lockdowns will not be sustainable. However, they are likely to continue using other measures such as masks, testing and tracing to mitigate transmission.

The UK and Sweden, more than most, have chosen to encourage mass infection while simultaneously denying vaccination to children. Sweden delayed vaccination of all children, before allowing vaccination for 16- to 18-year-olds in August 2021, followed by 12- to 15-year-olds. At the time of writing, they are still not vaccinating children aged between 5 and 11 years. In a misguided and unscientific population experiment, Sweden even explicitly referred to 'using' children to get them infected and create an 'immunity wall' for adults. The UK long argued against vaccination of children and followed similar timelines to Sweden for 12- to 18-year-olds. During 2020, the UK had 25 deaths in children under 18 years. By May 2021, according to data released under a freedom of information request, the UK had 37 child deaths – many more than the annual deaths caused by several other diseases that children are vaccinated against.

Death data by age (specifically child deaths) are not routinely reported in the UK. There has been a trickle of information following freedom of information requests, but no easily available aggregated data on exactly how many children have died in the UK. This is convenient, as releasing such data would make it easy to calculate how many lives of children could have been saved by vaccination. If measles, pertussis or tetanus were to cause 25 deaths a year in children, paediatricians would have been in uproar. Not only have preventable deaths been minimised, but chronic illness caused by COVID-19 has been ignored. In mid-February 2022, with crippling school outbreaks and ongoing child deaths, vaccination for 5- to 11-year-olds was reluctantly introduced. Then the UK dropped all mitigations and allowed people to attend work or school while infected. Sweden, too, has obfuscated data to minimise COVID, as revealed in a *Nature* publication. But at least 22 children have died due to the virus from a total population of 10 million people. The US still reports deaths data by age, and as of 26 March 2022 had 355 deaths from COVID-19 in children up to four years, and 737 deaths in children between five and 18. There is a far higher number of survivors with long COVID, including neurological effects, and there is no measurement of this in routine disease surveillance (among those countries, that is, who still count cases).

Encephalitis

There is a long list of other pathogens that harm the brain, including Japanese Encephalitis, which was declared a Communicable Disease Incident of National Significance on 4 March 2022. Japanese Encephalitis is a virus spread by Culex species mosquitoes. Until 2021 it had only ever been identified in the Torres Strait Islands of Australia and never on the mainland. The main vector of Japanese Encephalitis is the *Culex tritaeniorhynchus* mosquito. Until 2020 this mosquito had never been found in Australia. It was detected in Darwin and Katherine in the Northern Territory in 2019. Perhaps the mosquito had been silently spreading south since then. Aided by a La Niña year, with heavy rainfall and winds, Japanese Encephalitis declared itself in February 2022, detected in piggeries in Victoria. Since then, 14 human infections have occurred in Victoria, New South Wales and even South Australia, none of which had ever previously seen Japanese Encephalitis. This is deeply concerning – how did Japanese Encephalitis, a disease unknown on the mainland of Australia, suddenly spring up in the southern parts of the country? The main theory is the virus has spread through a different Culex mosquito, *Culex annulirostris*, which is present across Australia and associated with other infections such as Ross River Virus, Barmah Forest Virus and Murray Valley Encephalitis, which is an Australian virus that causes a serious encephalitis that is fatal in 15–30 per cent of cases.

Most people with Japanese Encephalitis get fever, headache and nausea or vomiting, and about 1 per cent develop encephalitis, with confusion, seizures and even coma. Up to 30 per cent who get encephalitis will die. At the time of writing, two deaths from the virus have occurred in Australia. The minimisation of COVID-19 is a striking contrast as tens of thousands of cases occurred every day during the BA.2 and then 4/5 variant surge, and hundreds a week continue to die of COVID-19. A 1 per cent risk of severe outcomes in kids is sneered at with COVID-19 as trivial, but Japanese Encephalitis is rightly declared a Communicable Disease Incident of National Significance with a 1 per cent rate of severe complications and vastly lower case numbers. There is a vaccine for Japanese Encephalitis, previously reserved for people in the Torres Strait. As we speak there are deliberations on broadening the criteria for vaccination.

The other cause of rapidly deteriorating and fatal brain damage is prions. These are neither bacteria nor viruses, but strange, misfolded proteins that can transfer their shape onto other proteins, thereby damaging them. Prion disease causes a fatal, degenerative encephalitis, referred to as spongiform encephalitis, usually decades after being consumed in contaminated meat. A universally fatal disease called Kuru was known among the Fore people of Papua New Guinea, peaking in the 1950s. It featured tremors, laughing fits, slurred speech, loss of coordination and, eventually, inability to move or even swallow, and death. It was found to be associated with cannibalism; specifically, the eating of the brain of the deceased. By the 1960s, the practice of cannibalism was stopped, but until 2010 Kuru persisted, due to the long incubation period of prion disease. Elsewhere in the world, the same disease is called Creutzfeldt-Jakob disease and occurs decades or years after consumption of contaminated meat. Mostly, cases are sporadic, but occasionally, epidemics do occur.

One of the most prominent epidemics of recent decades was the outbreak in the UK of 'Mad Cow Disease' in the mid-1980s. The outbreak was due to unnatural feeding practices, where farmed cattle were fed meat and bone meal made from dead sheep. Sheep themselves have their own prion disease, called scrapie. Scrapie has similar symptoms to human spongiform encephalopathies, such as loss of balance and difficulty walking. Through the practice of feeding meat to herbivores, British cattle became infected with the prion and therefore became contaminated. The cattle developed spongiform encephalopathy, and over four million cows were subsequently slaughtered. At least 177 humans who had consumed the beef developed variant Creutzfeldt-Jakob disease and died. The cattle outbreak began around 1985, and spread to other animals, including a domestic cat. The first human cases occurred in 1994, with a 19-year-old man dying in 1995. This was surprising because the disease is usually seen in older people. Following the cattle outbreak in the 1980s the US placed a ban on importing British beef which resulted in considerable losses for the UK beef industry. In 1990, in an effort to increase consumer confidence and prove that British beef was safe, the Agriculture Minister John Gummer fed his daughter a beef burger on national TV. It took until 1996 for the UK government to admit the outbreak was caused by eating prion-contaminated British beef. By that time the EU had joined the US in banning British beef. The British government's response is another interesting case study of economic priorities taking precedence over protecting health.

So, whilst fictional depictions of mass infection with brain-eating viruses are often far-fetched, there is certainly some scientific basis for this fear. Many pathogens can indeed alter the human brain, causing frightening symptoms and even death. Because of the sheer scale of infection globally, SARS-CoV-2 will have the greatest impact, causing substantial cases of encephalitis. To illustrate this point, a more severe but less transmissible pathogen-like variant Creutzfeldt-Jakob disease causes about 350 cases in the US each year. By contrast, SARS-CoV-2, which is highly transmissible, has seen at least 80 million cases in the US in the space of two years. The estimated rate of SARS-CoV-2 encephalitis is 0.25 per cent. That would equate to 100 000 cases of encephalitis a year. These would not be as universally fatal as Creutzfeldt-Jakob disease is, but the general point stands that cases of encephalitis resulting from SARS-CoV-2 stand to be vastly greater than for any other known pathogen. Even West Nile Virus, another important cause of encephalitis in the US, has a small burden of disease comparatively, with 1855 neuro-invasive cases in the US in 2021.

In terms of its effects on the brain, we do not yet know how SARS-CoV-2 will play out over the long term. But we do know that the virus can persist in the brain after the initial infection. The authors of a 2021 preprint study from the US National Institutes of Health that demonstrated this viral persistence drew a parallel with measles and asked whether we may see delayed encephalitis like the 'long measles' syndrome of subacute sclerosing panencephalitis. We can transplant hearts, lungs, kidneys and livers. But we can't do the same for brains. As a result, some illness that affects the brain may be irreversible. This makes prevention of SARS-CoV-2 critical.

13

THE BIOLOGICAL SNIPER WEAPON

SOME BIOLOGICAL OR CHEMICAL ATTACKS MAY BE targeted at an individual rather than a population, to disable or remove critical targets or high-value individuals. In 2017 North Korean leader Kim Jong-Un's half-brother Kim Jong-Nam was assassinated at Kuala Lumpur International Airport with VX nerve agent, a lethal synthetic chemical compound. Two women attacked Kim Jong-Nam with the substance – one splashed liquid in his face and the other covered his face with a cloth laced in VX. Within 15 minutes he was dead. It was followed in 2018 with the poisoning of Russian double agent Sergei Skripal and his daughter Yulia in Salisbury, UK, with the nerve agent Novichok. Novichok is a next-generation nerve agent thought to be five to eight times more potent than VX and designed to evade the Chemical Weapons Convention. It was Bellingcat, an investigative journalism site in the Netherlands, that identified those responsible for the poisoning attack of the Skripals. The culprits were named as Colonel Anatoliy Chepiga, who had been awarded the title of Hero of the Russian Federation in 2014 (possibly for his activities in Crimea or Ukraine), and Dr Alexander

Yevgenyevich Mishkin, a military doctor. They also identified the involvement of a third man, Denis Sergeev.

Bellingcat investigated the aliases provided by UK authorities, retraced their steps, identified the perpetrators and showed how brazen the attack had been – right in the heart of Salisbury, a military town. Travelling under aliases and supposedly working for a sports nutrition company, Colonel Chepiga and Dr Mishkin appear to have taken a commercial flight to London and then travelled on to Salisbury to carry out the attack. Bellingcat also located the site of a clandestine facility in Russia that was likely manufacturing the nerve agent, in a building posing as a sports nutrition facility. In Salisbury, Novichok was allegedly sprayed on the front door handle of the Skripals' home – enough to poison not only Skripal and his daughter, but also a police officer wearing full PPE who responded after the incident had been determined to be a chemical attack. Since Salisbury is a military town, the doctors at the emergency room of Salisbury Hospital knew how to recognise nerve agent poisoning.

Nerve agents act in the same way as organophosphates, by inhibiting the enzyme that breaks down acetylcholine, an organic chemical that functions as a neurotransmitter in the brain and body. This inhibition results in accumulation of acetylcholine in the nervous system, causing excess salivation and respiratory secretions, wheezing, shortness of breath, diarrhoea, vomiting, sweating, headaches, altered mental state, weakness, palpitations and paralysis. Pinpoint pupils are indicative in people presenting with these symptoms. If the drug atropine is given immediately, it breaks down acetylcholine and reverses the effects. Had the victims of the attack been admitted to another hospital, the doctors may not have diagnosed the syndrome soon enough to save Skripal and his daughter with atropine.

Three months after the attack on the Skripals, a discarded perfume bottle allegedly containing the substance was found in a park seven miles away by a local man, Charlie Rowley. He took it home to his partner, Dawn Sturgess, who sprayed it on her wrists, became ill immediately, and died a month later in hospital. Experts estimate the bottle contained enough Novichok to kill thousands of people. A massive environmental decontamination effort ensued for this previously unknown nerve agent that was invisible and impossible to identify. In 2020 Novichok was used again, this time against the Russian opposition leader Alexei Navalny. Navalny managed to survive the poisoning.

Chemical weapons used to target individuals are more obvious than biological weapons as they carry a specific signature and appear clearly unnatural. Sometimes targeted attacks are carried out to make a statement, and perhaps issue a warning, as seems to have been the case with the Skripal attack, so chemical weapons are a more obvious choice. On the other hand, a stealth attack may be preferred when a bad actor does not wish to attract attention, but wants an enemy disposed of.

Precision harm and high-value individuals

After years of researching biosecurity and observing the rise of precision medicine (where medicine is custom designed for individuals based on their unique genetic make-up), I realised there was a flip side to this: a dual use for intentional harm. I coined the term 'precision harm' to describe targeted harm enabled by technology which exploits individual illness or genetic vulnerability.

Like most people, high-value individuals have a large amount of digital information available about them, including banking, financial, administrative, human resources and personal health data. The value of electronic personal health records and the increasing targeting of hospitals in cyberattacks are well recognised. Electronic health records, big data, data linkage and cyberattacks enable health profiling which can be used to plan precision attacks. In the era of precision medicine, personal genetic data (especially any genetic material submitted for medical testing or for commercial genealogical services like Ancestry.com) could also be included in a high-value individual's profile, creating a new level of vulnerability to 'precision harm'. Digitisation of personal health information, including genomic information, greatly facilitates this threat.

In 2014 a catastrophic hack was carried out on the US Office of Personnel Management (OPM), the agency responsible for 20 million US federal employees. At the same time, the largest health insurance provider for federal employees in the US, Anthem Health, was also hacked. If the perpetrators of the two incidents were the same or connected, they could have linked the data from OPM to Anthem's sensitive health information, providing a blueprint for precision harm against high-value federal employees – a judge with diabetes, for example. Or a politician with heart disease who wears a pacemaker digitally linked to the cloud so her cardiologist can monitor her heart. Indeed, in anticipation of a possible hacking attack, former US Vice President Dick Cheney had the wireless function disabled in his implanted defibrillator and heart pump precisely to avoid such a situation. Today there are many internet-connected medical devices, including pacemakers, insulin pumps and drug delivery systems. Whilst these devices do make clinical care easier, there is the potential to hack them to cause harm.

If a stealth attack is designed to appear 'natural', it would not involve agents such as VX, anthrax or smallpox, as these would signal foul play. In this case, an attack may involve a garden variety pathogen (for example, salmonella, influenza) but with high-dose exposure which would ensure that the illness starts earlier, is more severe, and more likely to be fatal. A person given a mega-dose of influenza may not present like someone with regular flu – they will be much sicker. One target may experience multiple or serial attacks, which manifest as abnormal frequency of infections, and more severe infections. Sometimes the goal may be incapacitation rather than death, especially when the target is a prominent individual. If the objective is to disable a high-value target (for example, a judge in an important organised crime trial), recurrent illness would achieve the aim. Recurrent severe infections in a single individual with normal immune function may signal that the person has been targeted with a biological weapon. In other cases, a targeted attack may involve an agent with a longer onset but universally fatal outcome, such as variant Creutzfeldt-Jakob disease. But if an unnaturally high dose were administered, even that disease would come on quickly.

The most vulnerable locations for high-value individuals are those places predictably frequented, such as the home, office or vehicle. Planned travel, aircraft, dining, and other scheduled events are similarly high risk. Long-term attacks, such as radiation exposure with the intent of causing cancer, would likely be perpetrated in the home, office or in a person's vehicle. A one-off attack, on the other hand, could be conducted in the same venues or in less frequented but more accessible places such as aircraft, hotels or restaurants. A point of extreme vulnerability is when a high-value individual is hospitalised for illness or surgery. There they might be fitted with an intravenous drip, which offers an easy portal for administering toxic agents either by substitution of prescribed drugs or by stealth while the person is asleep or incapacitated. Research on insider threat suggests that, if required, it would be relatively easy to bribe, blackmail or otherwise coerce a colluder in any of these settings, including hospitals and the health system.

A range of attacks could be made on high-value individuals, with either long-term or acute effects, including infection, carcinogenesis or exposure to radiation or toxins which could affect various organ systems causing syndromes such as cardiac arrest, acute or chronic respiratory disease, neurological or gastrointestinal illness. Long-term exposure to gamma radiation could be achieved by placing a radioactive object in a site where the victim would be repeatedly or constantly exposed, such as the home, car or office. Long-term exposure to pulmonary irritants, through ventilation or air-conditioning systems in the home, office or car, could produce lung damage. One-off attacks could be caused by exposure to a biological or chemical agent through direct contact (touch), inhalation, injection or sexual transmission, or through ingestion of deliberately contaminated food, beverages or medications. In Chicago in

1984, Tylenol (paracetamol) medication was contaminated with cyanide, killing seven people. This incident led to the development of blister packaging for medicines. Medication security remains a problem, with continuing examples of tampering. In 2017, for instance, pills within blister packs of Valium were replaced with another drug by an employee in a factory in Australia, highlighting that tampering can occur anywhere from the factory to the shelf to the home.

Medication tampering

In 2014, in a rebel-held part of Syria, between 15 and 34 infants died and at least 50 more became ill following a WHO measles vaccination campaign. The measles vaccine comes in two vials: the active component in one vial as a lyophilised powder, with a diluent (usually saline) in the other. The two must be mixed together before administration. In this case, however, the diluent, packed and shipped with the vaccine for the WHO campaign, was not saline, but atracurium – an anaesthetic agent related to curare that paralyses all muscles, including the diaphragm, resulting in an inability to breathe. The infants injected with this vaccine suffered a terrible death. The official WHO investigation deemed the incident accidental, without providing any explanations for how the vials of atracurium had ended up being packaged with the measles vaccine and shipped to Syria. Given that the incident occurred in a rebel-held part of Syria at the height of the conflict with Islamic State, deliberate tampering cannot be ruled out.

The 2014 measles vaccine case in Syria highlights the possibility of vaccines being used as weapons, not only to kill people but to cause loss of trust in vaccination programs and to incite anti-vaccination sentiment. Four years later, the very same lethal incident occurred in Samoa, resulting in the death of two infants. How on earth could exactly the same, bizarre and unlikely accident occur twice, in two countries so far apart, especially after a supposed investigation into the Syrian incident, which should have ensured this never occurred again? In Samoa, it resulted in a loss of trust in vaccines, fuelled anti-vaccination sentiment, and led to a drop in vaccination rates. In 2019 a serious epidemic of measles broke out in Samoa with over 5000 cases and 70 deaths. It was so severe that a national emergency was declared. Measles causes an immune paresis for two to three years after infection, during which time people are more susceptible to any other infection. The timing of the measles epidemic therefore left Samoa very vulnerable to the COVID-19 pandemic.

Another notable case of a vaccine being used for a targeted assassination occurred during the US campaign to locate Osama Bin Laden. In this case, a vaccination campaign was used as cover to get DNA samples from children, in a bid to identify Bin Laden's offspring and in turn, his exact hiding location. In 2010, the CIA organised a fake vaccination drive in Abbottabad without notifying the Pakistani government, and co-opted a Pakistani doctor, Shakil Afridi, to help. Some have suggested that USAID and Save the Children were also involved. Children were offered free hepatitis B vaccines, beginning in a poor area in the outskirts of Abbottabad. It appears they only gave one out of a three-dose schedule to these children before moving on to the more affluent Bilal Town, where Bin Laden was suspected to reside. It is clearly unethical to give children a single dose of hepatitis B vaccine and not to provide the subsequent required doses. How DNA from vaccinated children in Bilal Town was obtained is unclear. Some speculate that nurses giving the vaccination may have drawn blood back into the syringe for a biological sample.

The scheme seems not to have helped identify Bin Laden's location, but other information obtained by Dr Afridi about the name and phone number of a resident in the Bilal Town compound, was enough to green light the operation leading to Bin Laden's assassination late in the evening of 2 May 2011. US SEAL Team Six completed their mission, spending less than 40 minutes on the ground. The Pakistani government was furious that the US had conducted an operation in their country without their knowledge. Pakistan's Inter-Services Intelligence agency arrested Afridi soon afterwards, and he remains in prison. When news of the use of the vaccines to find Bin Laden became public, there was a backlash among local Pakistanis, who had already been hesitant about vaccines. They lost all trust in vaccine programs and became fearful of vaccines being used to harm their children. WHO has been close to eradicating polio, except for three countries where it has not been eliminated – Nigeria, Afghanistan and Pakistan. Since 2012, multiple polio vaccination teams, including nurses and doctors, have been assassinated in Pakistan. It is estimated that at least 70 polio workers have been shot dead since then, with the Taliban claiming responsibility for some of these attacks.

A smorgasbord of options

The mode of an attack depends on the route of infection, which varies according to the type of pathogen. Pathogens such as Ebola are transmitted through direct contact. In an Ebola attack the pathogen could be sprayed or smeared onto a surface such as a door handle, steering wheel, coffee cup, computer mouse, keyboard and other commonly touched surfaces. Pathogens like smallpox, influenza, legionella, SARS-CoV-2 and tuberculosis, meanwhile, are airborne. In an attack using these pathogens in closed spaces such as the office, home, vehicle, aeroplane or hotel room, the air-conditioning systems or other ventilation systems may be used. Some pathogens such as influenza or SARS-CoV-2 can remain airborne hours after release, so could be sprayed in a room prior to the target entering. Viruses such as HIV and hepatitis B and C can be transmitted through needles, medical instruments, blood transfusions or sexual intercourse. From a perpetrator's perspective, a high-value individual visiting a dentist or a doctor for an invasive medical procedure would represent an opportunity for such an attack. Relevant equipment could be contaminated with or without the knowledge and collusion of the health practitioner. Bribery or blackmail could be used to co-opt collusion in such an attack.

Cancers, especially blood or bone marrow cancers, can be induced by radiation exposure and also by gene editing technology. Low-dose radiation exposure would likely be perpetrated in the office, home or vehicle, all areas where a high-value individual may be expected to spend long periods of time. Gene editing and integration of carcinogenic genes into the host genome could be achieved through the food chain, viral vectors, vaccines or medical interventions. In a recent development, malware has been developed which can hack and modify CT and MRI scans to falsely show cancer, or to falsely remove evidence of cancer. This means that a high-value individual with cancer could be wrongly cleared, or one without cancer could be sent down a pathway of unwarranted invasive surgery and chemotherapy.

A range of toxins and medicines are capable of causing sudden death through cardiac arrest. These can be incorporated into commonly consumed food or drink, prescription medicines, vitamins and non-prescription medications, or be planted during a meal out as a one-off attack. For foods or medicines, contamination or substitution may occur anywhere from the factory to the shop shelf to the home. Drugs able to induce cardiac arrest include potassium, prescription medications such as adrenaline (epinephrine), amiodarone and dihydropyridines, as well as illegal drugs such as cocaine and amphetamines. Insulin can also be used to cause sudden death, but requires injection, so would be more likely in hospital or in another situation where a high-value individual would concede to receiving an injection. In the US in the 1980s Claus Von Bülow was tried for two attempts to murder his wife with insulin. After initially being found guilty, he appealed and was acquitted.

High-value individuals might be targeted by drugs, infections or toxins that affect the brain or nerves. This includes infections such as prions as well as heavy metal poisoning including bismuth, mercury and magnesium. This can cause a range of neurological syndromes, acute delirium or dementia.

Several medication groups as well as illicit drugs can cause acute confusion or longer term dementia-like symptoms. Proton pump inhibitors and statins have been associated with memory loss. In addition, perpetrators may have developed unknown/experimental agents designed to alter cognition. Those with asthma or chronic lung disease may be targeted for exacerbation of their illness by means of inhaled lung irritants, chemicals and allergens. Asbestos fibres could be used to induce lung cancer and mesothelioma, although onset of cancer is typically many years after exposure.

Toxins or poisons such as ricin can be used to repeatedly poison a high-value individual, making them acutely or chronically ill. Dining out, travel, office, home and aircraft meals are all a risk. The supermarket shelf, the pantry and commonly consumed grocery items may also be a risk. Some toxins, such as ricin, are difficult to identify or test for, and others may be prescription medicines in high doses. A notorious case featured in *Bitter Harvest*, a book by true crime writer Ann Rule, was that of Debora Green, an American physician who repeatedly poisoned her husband, also a doctor, with ricin. This began when he tried to leave her, so she made him too sick to leave. Ricin cannot be traced in the body, unless you test for antibodies to it after repeated exposure, so doctors could not diagnose what kept making the husband ill. However, after she burnt down her house and killed two of her three children, police found the evidence. Ricin is made from castor beans, and they found castor beans in the house, and traced a purchase she had made for the beans.

Bio-self-defence

High-value individuals should be versed in protecting themselves against the array of methods that can be used against
them. They should receive training in basic infection control
techniques, such as wiping down commonly touched surfaces,
and be equipped with basic sanitisation equipment like alcohol
hand gels, along with disposable respirators when travelling
and for protection against airborne diseases like COVID-19.
A biological attack with an abnormally high exposure dose
may overwhelm the immune system, so high-value individuals
should be vaccinated against vaccine-preventable diseases such
as COVID-19, influenza, pneumococcal disease, meningococcal
disease, hepatitis A and B, pertussis, measles and other
scheduled vaccines. Even if vaccination does not prevent illness,
it may help avert severe complications and death. In addition, in
order to effectively and efficiently identify any attacks on their
staff members, organisations such as police, military or other
government agencies should routinely collect work health and
safety data on rates of illness (cancer, heart attacks, infections)
and deaths which can then be compared with general population
rates. Long-term trends or identification of occupational
cancer or infection clusters can help identify emerging threats.
Attacks, including low-dose radiation exposure, that are carried
out slowly and over a long period of time are more insidious.
Without ongoing monitoring, deliberate harm can be more
difficult to identify because symptoms and effects of low-dose
exposure might simulate apparently natural causes.

In order to mitigate against an attack, high-value individuals and their family members should take additional

precautions such as not using credit cards or loyalty reward cards when grocery shopping, as the data collected from these enable a digital profile to be created of frequently consumed items. Businesses and manufacturers, especially of food products and pharmaceuticals, have a responsibility to minimise the risk of tampering.

Healthcare institutions such as hospitals represent particularly high-risk settings for high-value individuals. Due to the ingrained belief that medical professionals would not harm their patients, hospitals and other places of treatment are ready potential environments for undetectable homicide of a high-value individual. Invasive procedures in hospitals, clinics or dental surgeries are also a risk. High-value individuals should avoid scheduling dental or medical appointments where minor invasive procedures are planned (involving needles, surgery or other breach of skin or mucous membranes), and instead try 'walk in' appointments, and avoid using phones or electronic communications. If hospitalised, protective measures should be applied at every step. This may include background checks on staff in the ward, arranging a shared (rather than single) room and having a carer present at all times.

High-value individuals should also be trained to question all medical procedures or treatments and to not give blind trust to a health professional. If a drug or IV is being administered, they should feel confident to ask what it is, and for what purpose, and who is administering it. In fact, everyone who is a patient should feel confident enough to question a nurse or doctor. It is difficult, including for me. I once had surgery and a nurse walked into my hospital room to change my dressing. He first emptied the bin, handling it without gloves, and then came straight to me to change my dressing. I watched him approach, horrified, and found it really difficult to summon up the courage to tell him to wash his hands. Just before he touched me, I said, 'Could you please wash your hands before changing my dressing?' and he obliged. No one else will care about your safety as much as you do, so when you are a patient, always ask questions if something does not seem right.

We think of bad actors using guns to assassinate enemies, but increasingly, homicide can be committed by stealth and may pass for natural death. Given the expanding use of biological weapons, high-value individuals today face an array of plausible biological threats. This necessitates the implementation of effective protection plans, both by high-value individuals themselves and by the organisations employing them. If hostile states, terrorist groups or other well-funded actors are using targeted chemical, biological and radiological methods of attack, then high-value individuals are easy targets, especially if they are sceptical of the possibility. If they are not made aware and prepared for the expanding threats they face, high-value individuals will be thrust into an uneven contest against well-equipped adversaries who attack by stealth and may never be detected.

14

A BIOLOGICAL WINTER

THE RELENTLESS MARCH OF TECHNOLOGY, THE denial of unnatural epidemics, the insider threat and the weak governance of biosecurity leads me to the conclusion that we are utterly unprepared. A 'biological winter' is a term I coined to describe something analogous to a nuclear winter. But it's a world full of engineered and synthetic pathogens that keep coming at a faster rate than we can design vaccines and drugs. Pandemics are big business, and there are many who profit from them. A biological winter is possible, and we may even be in the midst of one. And chances are we will not even recognise what is in plain sight, or that it will be hidden from us by the use of information warfare.

In the examples I have described throughout this book, cognitive dissonance, denial and information warfare are recurring themes. Many examples, from the Rajneesh attack to Sverdlovsk, the 'Russian flu', Amerithrax and the COVID-19 pandemic, demonstrate scientists' vested interests in denying the possibility of unnatural epidemics. Other examples such as Operation Sea Spray and the DICE Trials illustrate the harm that may be inflicted by governments on their own soil.

To this day, many apologists deny the use of smallpox against Indigenous people in the US and Australia, because it does not fit with their perception of the foundation of these countries.

The 1984 Rajneesh bioterrorist attack saw truth shouted down by public health authorities, a confession not believed, and the facts covered up for over a decade. In this case there was no government or insider scientist involvement (so no obvious motive for cover-ups), yet the same recalcitrance in considering unnatural origin was seen. Why is it that the American people were not told immediately about the Rajneesh attack? Perhaps if the general population had been informed, other incidents perpetrated by the Rajneeshis would have been uncovered and justice may have been served. The riveting 2018 Netflix documentary about the Rajneesh cult *Wild Wild Country* detailed many facts that were not widely known before, including the cult members' regular use of poisoned chocolates and food items to target individuals. A cult like that operating today would have access to far more sophisticated methods for precision harm and new ways to assassinate their targets.

As we have seen throughout the COVID-19 pandemic, it is common for governments and experts to withhold truth from ordinary people for fear of 'causing panic'. Perhaps this explains the silence around the Rajneesh attack. It may explain why we were brainwashed with repeated messages about SARS-CoV-2 being trivial in children; or that the virus is not airborne. The word 'airborne' fills many experts with dread, probably because it requires organisational and systemic change like providing safe indoor air, ventilation systems and N95 respirators. By comparison, it is easier to shift blame onto individuals for personal failings, like forgetting to wash their hands. There are also darker roots to this denial of airborne transmission of COVID-19, stemming from the 2003 SARS epidemic. Back then, too, the argument about large droplets as opposed to airborne transmission raged. And it took a real human toll, as a tale of two cities in Canada reveals.

During the 2003 SARS outbreak, in Vancouver N95 respirators were provided for health workers, and in Toronto, only surgical masks. Experts in Toronto insisted SARS was not airborne and that a surgical mask was enough. Both cities had their first SARS case at the same time, but Vancouver prevented the spread and saved lives, while Toronto had a large outbreak resulting in 330 infections and 44 deaths – almost all of them in people who visited, were patients or worked in a particular hospital, or contacts of the same. While the numbers pale in comparison to SARS-CoV-2, the SARS Commission in 2006 was damning in its findings that Toronto authorities failed to be precautionary and protect their health workers. During COVID-19, that lesson was forgotten and the same mistake repeated. Some of my colleagues in North America think the reason for the denial of SARS-CoV-2 being airborne was related to the Toronto outbreak of SARS – an escalating commitment to a failing proposition that began all the way back in 2003. Tragically, this time the death toll was far greater.

The potential of biological research for good or for harm, the insider threat in laboratories, and the rise of DIY science are all truths supported by real examples. Yet the perspective of science and health experts is starkly different from that of police and intelligence agencies who may investigate biological events. When a biological incident occurs, scientists are the first people that law enforcement and intelligence agencies turn to.

During Sverdlovsk, US intelligence agencies demonstrated considerable capabilities, analysing the incident using satellite and other intelligence they gathered, and correctly concluded that it was an unnatural outbreak – only to be overruled by the scientific experts. During COVID, intelligence made public by NBC News showed substantial changes in mobile phone activity and traffic patterns around the Wuhan Institute of Virology from 11 October through November 2019, with road closures and a substantial drop in human activity. They found no traffic in the area around the institute between 14 and 19 October. Based on these and additional open-source data, US intelligence agencies estimated a major event at the Wuhan Institute of Virology around this time. Yet they agreed with the advice of virologists that the cause of the pandemic was natural. They may have been influenced by a political agenda.

Professor Jeffrey Sachs from Columbia University is President of the UN Sustainable Development Solutions Network and chair of the Lancet COVID-19 commission. He disbanded the taskforce looking into the origins of SARS-CoV-2 when he became aware it was beset with conflicts of interest. His request to the National Institutes of Health for details of their research program on SARS-like virus research resulted in a completely redacted document being provided – only the cover page was unredacted. He concluded that the reason the US government was withholding data was that it may shed light on the origins of SARS-CoV-2. In an interview for *Current Affairs*, Professor Sachs explains that if SARS-CoV-2 emerged from a lab leak, the involvement of the US in funding gain-of-function research at the Wuhan Institute of Virology may be covered up because 'the implications are huge. Imagine if this came out of a lab. And we have, by some estimates, about 18 million dead worldwide from this … Well, the implications of that – the ethical, the moral, the geopolitical – everything is enormous.' Sachs was immediately attacked on social media by a group of virologists. This is the recurring tug-of-war between narrative and counter-narrative, seen during so many past examples and through the COVID pandemic.

In case after case – from the 1979 Sverdlovsk anthrax leak in the Soviet Union, to the 1984 Rajneesh salmonella attack in the US – health experts have demonstrated an inability or unwillingness to consider unnatural origins of epidemics. They have used extreme information warfare to create a narrative to support their case, best illustrated by the 1977 'Russian flu' pandemic. Major incidents, including testing of bioweapons in the UK and the US on their own people, have been followed by lengthy silences and cover-ups. Meanwhile technology has been racing ahead, becoming cheaper and more accessible. War, terrorism and conflict also continue in the world, with the advent of powerful non-state actors across the globe. Today, the array of actors with access to technology is far greater than was ever imagined even 20 years ago. Biological technology that was unthinkable back then is now very much a reality and, moreover, much more affordable.

Criminals and terrorists who previously worked in isolation can now connect in extensive global communities on the dark web, with new ways of colluding when there are mutual interests. Intelligence agencies have recognised there is a coalescence and collaboration between organised crime, terrorists, insiders and other groups, but it is difficult to identify an insider networking on the dark web. One of the earliest dark web markets was Silk Road. Started in 2013, Silk Road was created as a site for the sale of illegal drugs, other contraband and even murder for hire, where people could order, pay in Bitcoin and have their purchase delivered using the regular mail. The site operated on the TOR, a secret browser developed by the US military for untraceable browsing and now an open-source tool. The creator of the site, Ross Ulbricht, was ultimately arrested and imprisoned for double life plus forty years without parole. But many other dark markets followed, including AlphaBay, a site which sold an array of contraband including stolen accounts and hacked data. AlphaBay was also linked to bomb threats made by an extremist group. AlphaBay was shut down in 2017. Mysteriously, the site's founder was discovered dead in his Bangkok prison cell soon after. In 2021 AlphaBay was relaunched with new rules, including banning of sales of certain items. Curiously, it also prohibits discussion about Russia, Belarus, Kazakhstan, Armenia and Kyrgyzstan – causing speculation that the site is connected with these countries. Numerous other sites for dark web sales exist which could be used for trade in biological weapons and precursors. These platforms mean perpetrators could communicate, plan, purchase equipment and materials, and conduct gain-of-function research without any official record of such activities.

DIY biology has emerged as a major global movement. While DIY is valuable for non-biological inventions, such as open-access methods for making air purifiers, when it comes to biological experiments it carries a risk because of contagion. Biohacker labs have been on the radar of law enforcement since 2009, when the FBI started sending agents to DIY biohacking conferences. Today there are over 80 known biohacker labs worldwide; three of them are in Australia. The same arguments made around the risks and benefits of gain-of-function research apply to biohacking. Proponents argue that the movement enables open sourcing of products as well as tools for biological research. Many run courses for beginners to help them get started in their own experiments. Others rent out space in their labs to citizen scientists. Some use CRISPR/Cas 9 to engineer bacteria. Such labs are meant to operate at the lowest level of biosafety (biosafety level 1, which entails implementing basic precautions including regular hand-washing and minimal protective equipment). More dangerous pathogens are prohibited in the DIY space, but there is no formal oversight of practices in these labs.

Those who caution about the biohacking movement warn that there is inadequate safety and ethics training, and in fact work that would not be approved by ethics committees in universities could be carried out in DIY labs. Democratising science is a core agenda of DIY biology, which implies science should not be restricted to people with formal qualifications, institutional infrastructure and oversight but rather should be available to all. Should the creative drive of individuals override all other concerns? Whilst genuine innovations have arisen from the movement, such as development of cheap COVID-19 tests and open-sourced insulin, pathogen experiments may harm others beyond the DIY lab because of contagion. The safety and rights of all people, not just biohackers, must be considered.

Super-humans

By 2011 the biohacker community was sufficiently organised to create voluntary codes of conduct. California banned the use of CRISPR/Cas9 in biohacking in 2019. This was prompted by Josiah Zayner, a scientist with NASA at one time, who livestreamed himself injecting CRISPR/Cas9 into his body in 2017. He was investigated for practising medicine without a licence in California. The cyborg movement is also growing, with biohackers experimenting with inserting machine parts into humans. Sydney-based Meow-Ludo Disco Gamma Meow Meow was convicted for travelling without a ticket when he surgically implanted his Opal travel card chip into his hand, but the conviction was later over-turned. In 1998 the United Nations adopted the Universal Declaration on the Human Genome and Human Rights, which bans human cloning. The faux outrage after He Jiankui's experiment lasted a millisecond before everyone jumped on the bandwagon for fear of being left behind. In 2021 the WHO released recommendations on human genome editing, signalling the gates are open to this technology. In 2022 the US Food and Drug Administration released a report on human gene therapy products incorporating human genome editing.

Human germline editing has a strong likelihood of becoming the next arms race. In 2019 the US Defense Advanced Research Projects Agency announced its intention to explore genetic editing of soldiers, signalling their belief that China was already doing such research. In 2021, the UK announced the creation of an £800 million Advanced Research and Invention Agency to carry out human genome editing research. Unlike in the US, the UK research will not be subject to freedom of information requests, so who knows what is happening behind closed doors?

There is no doubt that human gene editing is part of our future. This, too, poses a dual-use dilemma. While human gene editing has the potential for positive outcomes like curing genetic disease, such research may also have many adverse consequences. The mainstream media makes much of athletics and competitive sport as a focus of ethical debate, but will we have a world where two classes of humans exist? The privileged and perfectly engineered ones, and an underclass of naturally born ones, as was depicted in the movie *Gattaca*. Human gene editing may also become the new arms race between superpowers, striving to create super soldiers and super citizens. In 2019, my colleague David Heslop and I addressed these issues in relation to the 2018 He Jiankui scandal. We pointed out that as a strategic military objective, in addition to creating strength and advantage for a country or group of people, gene technology could be used to incapacitate or weaken an enemy population. In parallel to major advances in pathogen biology, the impact of a biological weapon may be amplified using human gene editing to cause immunosuppression in a population.

It will be interesting to see how the COVID-19 pandemic plays into the geopolitical future of the US and China, each pursuing human genome engineering, but with very different approaches to preventing COVID. When they gave up on COVID control and opened the gates to mass COVID-related chronic disease and disability, the health folk in the US clearly didn't speak to the military folk who were aiming for the exact opposite with their engineered super-humans. In China, however, both agendas are aligned for now.

Cascading failures

Pandemics cause cascading failures in society. Weaknesses in the existing response framework have been laid bare during the COVID-19 pandemic. The collapse of health systems and supply chains was a weak point in society, even in wealthy countries. In pandemic plans, critical infrastructure, such as banking, power, water, food and telecommunications, are often assumed to be functioning uninterrupted. Indeed, core elements of the response plan even require these systems to be working. Without power, for example, vaccine fridges would not function and vaccines would be wasted. At the start of the COVID-19 pandemic in early 2020 when only China was badly affected, there was a worldwide shortage of certain medicines and medical supplies which depended on supply chains from China. Even large pharmaceutical manufacturing plants in India which supply major drug companies around the world could not produce drugs because they relied on precursors from China.

The fragility of the global just-in-time economy and reliance on globalised supply chains was exposed right at the start of the COVID-19 pandemic. The lesson was learnt during the 2009 influenza pandemic, when many countries established domestic manufacturing of essential medical supplies – but it had all been whittled away when COVID struck. Investing in and supporting domestic capacity in the manufacturing of essential medical supplies such as masks, PPE and medicines is an important pillar of preparedness. During the Omicron wave in 2022, supply chains were disrupted again, but this time due to mass workplace absenteeism. Normally, during business-as-usual, about 2 to 5 per cent of employees will be absent from work. During the Omicron wave absenteeism rates rose to 20–30 per cent. This crippled supermarket chains' ability to move products from warehouses to retail outlets. Farms could not harvest and transport their produce, resulting in the dumping of large quantities of rotting produce. There was no chicken on supermarket shelves – this was different from the panic-buying of toilet paper, which was demand-driven. This was a true supply crisis driven by massive numbers of people being too sick to work. In the UK, the genius solution was to remove requirements for infected people to isolate, thus allowing infected people to attend work and infect their colleagues. Workforce shortages would be accordingly fixed for about three days, before making absences even worse. The burden of long COVID has further impacted the workforce, with many unable to work with the debilitating long-term effects of the disease.

The same governments who decided we could 'live with' Omicron, failed to plan for the collapse of the workforce or for adequate COVID-19 testing infrastructure. In 2021, many people became RAT (rapid antigen test) millionaires by winning contracts to import RATs, but on the streets of Sydney during the surge in Omicron, no one could actually lay their hands on a RAT. Simultaneously, state governments across Australia, unable to cope with demand for testing, opted to restrict PCR testing instead of expanding it. Up until that point PCR testing had been freely available. At the same time, a glacial pace of procuring vaccines for children and an equally slow vaccine advisory committee, forced children back to school in 2022 unvaccinated. How hard would it have been to plan to have schoolchildren vaccinated for the start of school in 2022? The same occurred with booster recommendations, made too late to protect people against oncoming pandemic waves. Vaccine policy-making in a pandemic should have more flexible and agile processes than in normal times, or the opportunity to protect people against pandemic waves will be missed. Yet this is rarely explicit in pandemic plans.

A slippery slope

Current pandemic planning assumes recognition of biothreat, but serious epidemics are occurring at a faster rate than ever before, with few questions asked. Historically, most unnatural outbreaks have not been recognised as such. Only smallpox, being eradicated, would be immediately recognised as unnatural – and if smallpox were to re-emerge during the current monkeypox epidemic, it may be initially camouflaged by monkeypox. A disaster medicine expert from the US pointed out on social media that it is NATO countries that are most affected in the 2022 epidemic of monkeypox. He asked the question 'could this be used as a smokescreen for a coming smallpox attack?' (that would not be recognised for a while because it would be assumed to be monkeypox). Such questions do need to be asked, and anyone who becomes outraged about such questions should be scrutinised for their motives.

Policy decisions about biosecurity should be made considering the worst-case scenario, not the best-case scenario. Take the euthanasia debate, for instance: many advocates are driven by a noble motive, to prevent suffering in those who choose to end their lives, and many have moving, personal examples to draw on. However, creating legislation based on a best-case scenario is dangerous – it risks being the slippery slope to involuntary euthanasia of frail, older people who are taking up hospital beds or considered a burden to society. This was exactly what happened in some countries during the COVID-19 pandemic, with many reports of involuntary euthanasia of infected elderly people, along with the active withholding of simple measures like oxygen, and in some cases even food and water. Human history does not give us much hope that euthanasia legislation will not be abused. Given that the greatest cost to hospitals and health systems is the illness of older people, in the all-consuming drive for efficiency seen in all health systems, it is quite possible that hastening the death of older people may become the norm. We have already been sensitised to thinking of some lives as expendable during the COVID-19 pandemic, so the stage is set for a slippery slope downhill. In contrast to euthanasia, failing to consider the worst-case scenario in biosecurity could threaten the very survival of the human species. So it is imperative that policy decisions are made with both the best- and worst-case scenarios in mind.

Can we bring ourselves to contemplate an unimaginable future where human genome editing and resurrection of extinct viruses is the norm? There are many parallels here to the response to climate change, which has similarly been characterised by denial. However, there is far greater community awareness of climate change than of biosecurity. The level of basic knowledge about biosecurity is generally low in the community. There is a tendency to see only one side of the coin of biological science – the positive side, and to be unaware of the many negative ways it could change our future. When it comes to weighing up the risks and benefits of biological technology, we should not take a blinkered approach by considering only the benefits to humanity. There are so many barriers to even acknowledging the risks of dual-use biology, the greatest being the conflict of interest by those most qualified to discuss it. Much of the debate about the benefits and risks of biological research has occurred within the scientific community, with the rights of scientists themselves being the driving concern while the broader community is kept in the dark. This is a threat that affects everyone, crosses national boundaries, and which cannot be effectively managed either in the traditional disciplinary silos or by individual nation states. Instead, it requires coordination, thought leadership and novel, cross-disciplinary, global solutions.

The solution to the existential threat we face in biosecurity will not come from the health or medical research sector for all the reasons I have outlined. The greatest hope is that intelligence and law enforcement agencies will step up and accelerate their preparedness. Judging by conversations I've had with people in law enforcement and intelligence, I am doubtful that these agencies have fully grasped the scope of the modern threat landscape. Or if they have, they may be hampered by political agendas. When I deliver presentations at police and intelligence conferences about red flags for planned biological attacks, it is clear they lack comprehensive intelligence gathering for signals of such threats. They do not have the required methods for threat detection for possible biological weapons, or if they do, cannot apply available methods successfully. The Rajneesh case is an example where the methods and expertise were available, but where cognitive dissonance created a complete roadblock to uncovering the truth. We cannot wait for human thinking and fallibility to catch up with advances in technology. When specialists cannot grasp the threat, or worse, deny it, it is even more important that the community is empowered.

We should avoid dumbing down information or keeping the public in the dark about momentous changes occurring around them. We should not infantilise the community using 'fear of causing panic' as a convenient excuse to sidestep truth. It is essential we engage the community as the most important stakeholder in biosecurity and be guided by their voices. It is imperative that all people should understand the scope of benefit and harm that is possible with biological technology, and how it may affect their future, the future of their children and of the planet. The democratisation of knowledge is more valuable than the democratisation of biology. If the community is empowered with knowledge, the lens of history can help us to avoid repeating the mistakes of the past.

REFERENCES

Preface   Corona dawn

MacIntyre CR, Heslop DJ, Nguyen P, Adam D, Trent M, and Gerber BJ. Pacific Eclipse – A Tabletop Exercise on Smallpox Pandemic Response. *Vaccine*. April 14, 2022, 40(17): 2478–83.

Chapter 1 Believe the unbelievable

Alibek K, and Handelman S. *Biohazard: The Chilling True Story of the Largest Covert Biological Weapons Program in the World – Told from the Inside by the Man Who Ran It*. Delta, New York, 1999.

Barnett A. Millions were in Germ War Tests. Much of Britain was Exposed to Bacteria Sprayed in Secret Trials. *The Guardian*. April 21, 2002, <www.theguardian.com/politics/2002/apr/21/uk.medicalscience>.

Dickerson K. Secret World War II Chemical Experiments Tested Troops by Race. *NPR*. June 2015, <www.npr.org/2015/06/22/415194765/u-s-troops-tested-by-race-in-secret-world-war-ii-chemical-experiments>.

Hammond P, and Carter G. Field Trials. In: *From Biological Warfare to Healthcare*. Palgrave Macmillan, London, 2001.

Keyes S. A Strange but True Tale of Voter Fraud and Bioterrorism. *The Atlantic*. June 10, 2014, <www.theatlantic.com/politics/archive/2014/06/a-strange-but-true-tale-of-voter-fraud-and-bioterrorism/372445/>.

Lois Bessho Commendation for Chemical Warfare Service, <www.documentcloud.org/documents/2107685-besshocommendation.html>.

MACE E PAI COVID 19 ANALYSIS Redacted. Contributed by *NBC News Investigations* (NBC). 2020, <www.documentcloud.org/documents/6884792-MACE-E-PAI-COVID-19-ANALYSIS-Redacted.html>.

National Security Archive. Volume V. Anthrax at Sverdlovsk, 1979. US Intelligence on the deadliest modern outbreak. *National Security Archive Electronic Briefing Book* No. 61, edited by Robert A. Wampler and Thomas S. Blanton, November 15, 2001, <https://nsarchive2.gwu.edu/NSAEBB/NSAEBB61/>.

References

Thompson H. In 1950, the U.S. Released a Bioweapon in San Francisco. *Smithsonian Magazine.* July 6, 2015, <www.smithsonianmag.com/smart-news/1950-us-released-bioweapon-san-francisco-180955819/>.

Török TJ, Tauxe RV, Wise RP, Livengood JR, Sokolow R, Mauvais S, et al. A Large Community Outbreak of Salmonellosis Caused by Intentional Contamination of Restaurant Salad Bars. *Journal of the American Medical Association.* 1997, 278(5): 389–95.

Chapter 2 Insider threat

Alberts B, and Fineberg HV. Letter about Thomas Butler from Presidents of the National Academy of Sciences and the Institute of Medicine. August 13, 2003, <https://sgp.fas.org/news/2003/08/nas081503.pdf>.

Bird C. Steven Hatfill's Strange Trip from Accused Terrorist to Medical Adventurer. *Newsweek Magazine.* June 18, 2014, <www.newsweek.com/2014/06/27/steven-hatfills-strange-trip-accused-terrorist-medical-adventurer-255295.html>.

Bunk R. *In the Heartland: A Story So Real, It Might Be True.* Page Publishing Inc., La Vergne, 2020.

Engelberg S. New Evidence Adds Doubt to FBI's Case Against Anthrax Suspect. *Propublica.* October 2011, <www.propublica.org/article/new-evidence-disputes-case-against-bruce-e-ivins>.

Freed D. The Wrong Man. *The Atlantic*, May 2010, <www.theatlantic.com/magazine/archive/2010/05/the-wrong-man/308019/>.

Hatch Rosenberg B. UQ Wire: Dr Rosenberg Analysis of Anthrax Attacks. *Scoop.* June 29, 2002, <www.scoop.co.nz/stories/HL0206/S00180/uq-wire-dr-rosenberg-analysis-of-anthrax-attacks.htm>.

Kane P. Sen. Leahy on Anthrax Case: 'It's Not Closed'. *Washington Post*, February 11, 2011, <www.washingtonpost.com/politics/sen-leahy-on-anthrax-case-its-not-closed/2011/02/16/ABGl7UH_story.html>.

Ling J. A Brilliant Scientist Was Mysteriously Fired from a Winnipeg Virus Lab. No One Knows Why. *Macleans.* February 2022, <www.macleans.ca/longforms/winnipeg-virus-lab-scientist/>.

Murray BE, Anderson KE, Arnold K, Bartlett JG, Carpenter CC, Falkow S, Hartman JT, Lehman T, Reid TW, Ryburn FM Jr, Sack RB, Struelens MJ, Young LS, and Greenough WB 3rd. Destroying the Life and Career of a Valued Physician-Scientist Who Tried to Protect Us from Plague: Was it Really Necessary? *Clinical Infectious Diseases.* June 1, 2005, 40(11): 1644–48.

National Academies of Sciences, Engineering, and Medicine. *Review of the Scientific Approaches Used During the FBI's Investigation of the 2001 Anthrax Letters.* The National Academies Press, Washington, DC, 2011.

Pauls K. Canadian Scientist Sent Deadly Viruses to Wuhan Lab Months before RCMP Asked to Investigate. *CBC News.* June 2020, <www.cbc.ca/news/canada/manitoba/canadian-scientist-sent-deadly-viruses-to-wuhan-lab-months-before-rcmp-asked-to-investigate-1.5609582>.

Chapter 3 Error, not terror

Andersen KG, Rambaut A, Lipkin WI, Holmes EC, and Garry RF. The Proximal Origin of SARS-CoV-2. *Nature Medicine*. April 2020, 26(4): 450–52.

Eban K. 'This Shouldn't Happen': Inside the Virus-Hunting Nonprofit at the Center of the Lab-Leak Controversy. *Vanity Fair*. April 2022, <www.vanityfair.com/news/2022/03/the-virus-hunting-nonprofit-at-the-center-of-the-lab-leak-controversy>.

Field M. A Lab Assistant Involved in COVID-19 Research in Taiwan Exposed 110 People After Becoming Infected at Work. *Bulletin of the Atomic Scientists*. January 2022, <https://thebulletin.org/2022/01/a-lab-assistant-involved-in-covid-19-research-in-taiwan-exposed-110-people-after-becoming-infected-at-work/#.YfLU1quiV1c.twitter>.

Fitzpatrick M. Review: The Cutter Incident: How America's First Polio Vaccine Led to a Growing Vaccine Crisis. *Journal of the Royal Society of Medicine*. 2006, 99(3):156.

Jacobsen R. Inside the Risky Bat-Virus Engineering that Links America to Wuhan. China Emulated US Techniques to Construct Novel Coronaviruses in Unsafe Conditions. *MIT Technology Report*. June 29, 2021, <www.technologyreview.com/2021/06/29/1027290/gain-of-function-risky-bat-virus-engineering-links-america-to-wuhan/>.

Klotz L. The Grave Risk of Lab-Created Potentially Pandemic Pathogens. *Bulletin of the Atomic Scientists*. September 2021, <https://thebulletin.org/2021/09/the-grave-risk-of-lab-created-potentially-pandemic-pathogens/>.

Laboratory-Acquired Infection (LAI) Database. *American Biological Safety Association*, <https://my.absa.org/LAI>.

Lina SMM, Kunasekaran MP, and Moa A. Brucellosis Outbreak in China, 2019. *Global Biosecurity*. 2021, 3(1).

MACE E PAI COVID 19 ANALYSIS Redacted. Contributed by *NBC News Investigations* (NBC). 2020, <www.documentcloud.org/documents/6884792-MACE-E-PAI-COVID-19-ANALYSIS-Redacted.html>.

Nathanson N, and Langmuir AD. The Cutter Incident. Poliomyelitis Following Formaldehyde-Inactivated Poliovirus Vaccination in the United States During the Spring of 1955. II. Relationship of Poliomyelitis to Cutter Vaccine. 1963. *American Journal of Epidemiol*ogy. 1995, 142(2):109–40; discussion 7–8.

Ridley M. Why Did Scientists Suppress the Lab-Leak Theory? *Spiked Online*. January 2022, <www.spiked-online.com/2022/01/12/why-did-scientists-suppress-the-lab-leak-theory/>.

Rozo M, and Gronvall GK. The Reemergent 1977 H1N1 Strain and the Gain-of-Function Debate. *mBio*. 2015, 6(4): e01013–15.

Young A, and Blake J. Near Misses at UNC Chapel Hill's High-Security Lab Illustrate Risk of Accidents with Coronaviruses. *Propublica*. August 2020,

References

<www.propublica.org/article/near-misses-at-unc-chapel-hills-high-security-lab-illustrate-risk-of-accidents-with-coronaviruses>.

Chapter 4 Dr Jekyll and Mr Hyde of biological research

Baric RS, and Sims AC. Humanized Mice Develop Coronavirus Respiratory Disease. *Proceedings of the National Academy of Sciences of the United States of America*. 2005, 102(23): 8073–74.

European Academies Scientific Advisory Committee. Gain of Function: Experimental Applications Relating to Potentially Pandemic Pathogens. *EASAC Policy Report*, October 27, 2015, <www.ae-info.org/attach/Acad_Main/News_Archive/Report on Gain of Function research/EASAC_GOF_Web_complete_centred.pdf>.

Fouchier R et al. Transmission Studies Resume for Avian Flu. *Science*. February 1, 2013. 339(6119): 520–21.

International Gene Synthesis Consortium, <https://genesynthesisconsortium.org/>.

Jackson R, and Ramshaw I. The Mousepox Experience. An interview with Ronald Jackson and Ian Ramshaw on Dual-Use Research. Interview by Michael J. Selgelid and Lorna Weir. *EMBO Reports*. 2010, 11(1): 18–24.

Jacobsen R. Inside the Risky Bat-Virus Engineering that Links America to Wuhan. China Emulated US Techniques to Construct Novel Coronaviruses in Unsafe Conditions. *MIT Technology Report*. June 29, 2021, <www.technologyreview.com/2021/06/29/1027290/gain-of-function-risky-bat-virus-engineering-links-america-to-wuhan/>.

MacIntyre CR. Biopreparedness in the Age of Genetically Engineered Pathogens and Open Access Science: An Urgent Need for a Paradigm Shift. *Military Medicine*. September 2015, 180(9): 943–49.

MacIntyre CR, Adam DC, Turner R, Chughtai AA, and Engells T. Public Awareness, Acceptability and Risk Perception About Infectious Diseases Dual-Use Research of Concern: A Cross-Sectional Survey. *BMJ Open*. 2020, 10(1): e029134.

Osterholm MT, and Henderson DA. Public Health and Biosecurity. Life Sciences at a Crossroads: Respiratory Transmissible H5N1. *Science*. 2012, 335(6070): 801–802.

Risk and Benefit Analysis of Gain of Function Research. *Gryphon Scientific LLC*. April 2016, <http://gryphonsci.wpengine.com/wp-content/uploads/2018/12/Risk-and-Benefit-Analysis-of-Gain-of-Function-Research-Final-Report-1.pdf>.

Chapter 5. Jurassic Park for viruses

Althoff KN, Schlueter DJ, Anton-Culver H, et al., on behalf of the All of Us Research Program. Antibodies to Severe Acute Respiratory Syndrome Coronavirus 2 (SARS-CoV-2) in All of Us Research Program Participants, 2 January to 18 March 2020. *Clinical Infectious Diseases*. February 15, 2022, 74(4): 584–90.

Carrat F, Figoni J, Henny J, Desenclos JC, Kab S, de Lamballerie X, and Zins M. Evidence of Early Circulation of SARS-Cov-2 in France: Findings from the Population-Based 'CONSTANCES' Cohort. *European Journal of Epidemiology*. February 2021, 36(2): 219–22.

Cello J, Paul AV, and Wimmer E. Chemical Synthesis of Poliovirus cDNA: Generation of Infectious Virus in the Absence of Natural Template. *Science*. August 9, 2002, 297(5583): 1016–18.

Chavarria-Miró G, Anfruns-Estrada E, Martínez-Velázquez A, et al. Time Evolution of Severe Acute Respiratory Syndrome Coronavirus 2 (SARS-CoV-2) in Wastewater during the First Pandemic Wave of COVID-19 in the Metropolitan Area of Barcelona, Spain. *Applied and Environmental Microbiology*. March 11, 2021, 87(7): e02750-20.

Chen X, Adam DC, Chughtai AA, Stelzer-Braid S, Scotch M, and MacIntyre CR. The Phylogeography of MERS-CoV in Hospital Outbreak-Associated Cases Compared to Sporadic Cases in Saudi Arabia. *Viruses*. May 14, 2020, 12(5): 540.

Chen X, Chughtai AA, and MacIntyre CR. Application of a Risk Analysis Tool to Middle East Respiratory Syndrome Coronavirus (MERS-CoV) Outbreak in Saudi Arabia. *Risk Analysis*. May 2020, 40(5): 915–25.

Chen X, Chughtai AA, Dyda A, and MacIntyre CR. Comparative Epidemiology of Middle East Respiratory Syndrome Coronavirus (MERS-CoV) in Saudi Arabia and South Korea. *Emerging Microbes & Infections*. June 7, 2017, 6(6): e51.

Dembek ZF, Pavlin JA, Siwek M, and Kortepeter MG. Epidemiology of Biowarfare and Bioterrorism. In: *Medical Aspects of Biological Warfare*. Department of the Army, Army Medical Department, The Borden Institute. Washington, DC, 2018.

Doucleff M. Scientists Race to Answer the Question: Will Vaccines Protect Us Against Omicron? *NPR*. December 2021, <www.npr.org/sections/goatsandsoda/2021/12/02/1060624669/scientists-race-to-answer-the-question-will-vaccines-protect-us-against-omicron>.

Gardner LM, and MacIntyre CR. Unanswered Questions about the Middle East Respiratory Syndrome Coronavirus (MERS-CoV). *BMC Research Notes*. June 11, 2014, 7: 358.

Gardner LM, Chughtai AA, and MacIntyre CR. Risk of Global Spread of Middle East Respiratory Syndrome Coronavirus (MERS-CoV) via the Air Transport Network. *Journal of Travel Medicine*. September 5, 2016, 23(6): taw063.

Gianotti R, Barberis M, Fellegara G, Galván-Casas C, and Gianotti E. COVID-19-related Dermatosis in November 2019: Could This Case Be Italy's Patient Zero? *British Journal of Dermatology*. May 2021, 184(5): 970–71.

Kpozehouen EB, Chen X, Zhu M, and Macintyre CR. Using Open-Source Intelligence to Detect Early Signals of COVID-19 in China, Descriptive Study. *JMIR Public Health and Surveillance*. 2020, 6(3): e18939.

Lerner S, Hvistendahl M, and Hibbett M. NIH Documents Provide New Evidence US-Funded Gain-Of-Function Research in Wuhan. *The Intercept*. September 2021, <https://theintercept.com/2021/09/09/covid-origins-gain-of-function-research/>.

MacIntyre CR. The Discrepant Epidemiology of Middle East Respiratory Syndrome Coronavirus (MERS-Cov). *Environment Systems & Decisions*. 2014, 34(3): 383–90.

Magnuson M. The 1918 Flu Killed 40 Million People. This Man Is Re-Creating the Virus. *Popular Mechanics*. October 2014, <www.popularmechanics.com/science/health/a12897/the-man-who-could-destroy-the-world-breakthrough-awards-2014/>.

McMullan LK, et al. Characterisation of Infectious Ebola Virus from the Ongoing Outbreak to Guide Response Activities in the Democratic Republic of the Congo: A Phylogenetic and In Vitro Analysis. *The Lancet Infectious Diseases*. 2019, 19(9): 1023–32.

Menachery VD, Yount BL Jr, Debbink K, Agnihothram S, Gralinski LE, Plante JA, Graham RL, Scobey T, Ge XY, Donaldson EF, Randell SH, Lanzavecchia A, Marasco WA, Shi ZL, and Baric RS. A SARS-like Cluster of Circulating Bat Coronaviruses Shows Potential for Human Emergence. *Nature Medicine*. 2015, 21(12): 1508–13.

Noyce RS, and Evans DH. Synthetic Horsepox Viruses and the Continuing Debate about Dual Use Research. *PLoS Pathog*. October 4, 2018, 14(10): e1007025.

Noyce RS, Lederman S, and Evans DH. Construction of an Infectious Horsepox Virus Vaccine from Chemically Synthesized DNA Fragments. *PLoS One*. 2018, 13(1): e0188453.

Osterholm MT, and Henderson DA. Public Health and Biosecurity. Life Sciences at a Crossroads: Respiratory Transmissible H5N1. *Science*. 2012, 335(6070): 801–802.

Piplani S, Singh PK, Winkler DA, and Petrovsky N. In Silico Comparison of SARS-CoV-2 Spike Protein-ACE2 Binding Affinities Across Species and Implications for Virus Origin. *Scientific Reports*. June 24, 2021, 11(1): 13063.

Qiu J. How China's 'Bat Woman' Hunted Down Viruses from SARS to the New Coronavirus. *Scientific American*. March 2020 (updated June 2020), <www.scientificamerican.com/article/how-chinas-bat-woman-hunted-down-viruses-from-sars-to-the-new-coronavirus1/>.

Rahalkar MC, and Bahulikar RA. Lethal Pneumonia Cases in Mojiang Miners (2012) and the Mineshaft Could Provide Important Clues to the Origin of SARS-CoV-2. *Front Public Health*. October 20, 2020, 8: 581569.

Reconstruction of the 1918 Influenza Pandemic Virus. Questions & Answers. *Centers for Disease Control and Prevention*, 2005, <www.cdc.gov/flu/about/qa/1918flupandemic.htm>.

Selgelid M, and Weir L. Reflections on the Synthetic Production of Poliovirus. *Bulletin of the Atomic Scientists.* June 2010, <https://journals.sagepub.com/doi/pdf/10.2968/066003001>.

Tumpey TM, Basler CF, Aguilar PV, Zeng H, Solórzano A, Swayne DE, Cox NJ, Katz JM, Taubenberger JK, Palese P, and García-Sastre A. Characterization of the Reconstructed 1918 Spanish Influenza Pandemic Virus. *Science.* October 7, 2005, 310(5745): 77–80.

Watanabe T, Zhong G, Russell CA, Nakajima N, Hatta M, Hanson A, McBride R, Burke DF, Takahashi K, Fukuyama S, Tomita Y, Maher EA, Watanabe S, Imai M, Neumann G, Hasegawa H, Paulson JC, Smith DJ, and Kawaoka Y. Circulating Avian Influenza Viruses Closely Related to the 1918 Virus Have Pandemic Potential. *Cell Host & Microbe.* June 11, 2014, 15(6): 692–705.

Chapter 6 The self-replicating weapon

D'Errico P. Jeffery Amherst and Smallpox Blankets: Lord Jeffery Amherst's letters Discussing Germ Warfare Against American Indians. 2001, <https://people.umass.edu/derrico/amherst/lord_jeff.html>.

Field M. US Official: Russian Invasion of Ukraine Risks Release of Dangerous Pathogens. *Bulletin of the Atomic Scientists.* February 2022, <https://thebulletin.org/2022/02/us-official-russian-invasion-of-ukraine-risks-release-of-dangerous-pathogens/>.

Gambardello JA. Social Media Posts Misrepresent U.S.-Ukraine Threat Reduction Program. *Factcheck.org.* March 2022, <www.factcheck.org/2022/03/social-media-posts-misrepresent-u-s-ukraine-threat-reduction-program/>.

Jackson R, and Ramshaw I. The Mousepox Experience. An interview with Ronald Jackson and Ian Ramshaw on Dual-Use Research. Interview by Michael J. Selgelid and Lorna Weir. *EMBO Reports.* 2010, 11(1): 18–24.

MacIntyre CR, Chughtai AA, Seale H, Richards GA, and Davidson PM. Respiratory Protection for Healthcare Workers Treating Ebola Virus Disease (EVD): Are Facemasks Sufficient to Meet Occupational Health and Safety Obligations? *International Journal of Nursing Studies.* November 2014, 51(11): 1421–26.

Martinez I. Rhodesian Anthrax: The Use of Bacteriological & Chemical Agents During the Liberation War of 1965–80. *McKinney Law,* University of Indiana, <https://mckinneylaw.iu.edu/iiclr/pdf/vol13p447.pdf>.

Orwellian Whitewash: CDC Deletes Faulty Ebola Guidelines Poster; Paging Ebola Czar! *Twitchy,* October 2014, <https://twitchy.com/jessem-34/2014/10/19/orwellian-whitewash-cdc-deletes-faulty-ebola-guidelines-poster-paging-ebola-czar/>.

Warren C. Was Sydney's Smallpox Outbreak of 1789 An Act of Biological Warfare Against Aboriginal Tribes? *ABC.* April 2014d, <www.abc.net.au/radionational/programs/ockhamsrazor/was-sydneys-smallpox-outbreak-an-act-of-biological-warfare/5395050>.

Wilson JM, Brediger W, Albright TP, and Smith-Gagen J. Reanalysis of the Anthrax Epidemic in Rhodesia, 1978–1984. *PeerJ*. November 10, 2016, 4: e2686.

Chapter 7 The spectre of smallpox

Alibek K, and Handelman S. *Biohazard: The Chilling True Story of the Largest Covert Biological Weapons Program in the World – Told from the Inside by the Man Who Ran It*. Delta, New York, 1999.

Bryner, J. Just 2 Labs in the World House Smallpox. The One in Russia Had an Explosion. *LiveScience*. September 18, 2019, <www.livescience.com/russia-lab-stores-smallpox-explosion-fire.html>.

Centers for Disease Control and Prevention. Thinking Outside the Cowpox: The Discovery of a Pox-Related Virus, <www.cdc.gov/ncezid/dhcpp/featured_stories/cowpox.html>.

Costantino V, Trent MJ, Sullivan JS, Kunasekaran MP, Gray R, and MacIntyre R. Serological Immunity to Smallpox in New South Wales, Australia. *Viruses*. 2020, 12(5).

MacIntyre C, Seccull A, Lane J, and Plant A. Development of a Risk-Priority Score for Category A Bioterrorism Agents as an Aid for Public Health Policy. *Military Medicine*. 2006, 171(7): 589–94.

MacIntyre CR. Reevaluating the Risk of Smallpox Reemergence. *Military Medicine*. August 14, 2020, 185(7–8): e952-e957.

MacIntyre CR, Costantino V, Chen X, et al. Influence of Population Immunosuppression and Past Vaccination on Smallpox Reemergence. *Emerging Infectious Diseases*. 2018, 24(4): 646–53.

MacIntyre CR, Das A, Chen X, Silva C, and Doolan C. Evidence of Long-Distance Aerial Convection of Variola Virus and Implications for Disease Control. *Viruses*. 2019, 12(1).

MacIntyre CR, Doolan C, and De Silva C. The Explosion at Vector: Hoping for the Best While Preparing for the Worst. *Global Biosecurity*. 2019, 1(1).

Nguyen PY, Ajisegiri WS, Costantino V, Chughtai AA, and MacIntyre CR. Reemergence of Human Monkeypox and Declining Population Immunity in the Context of Urbanization, Nigeria, 2017–2020. *Emerging Infectious Diseases*. April 2021, 27(4): 1007–14.

Noyce RS, Lederman S, and Evans DH. Construction of an Infectious Horsepox Virus Vaccine from Chemically Synthesized DNA Fragments. *PLoS One*. 2018, 13(1): e0188453.

Roth A. Blast Sparks Fire at Russian Laboratory Housing Smallpox Virus. *The Guardian*. September 17, 2019, <www.theguardian.com/world/2019/sep/17/blast-sparks-fire-at-russian-laboratory-housing-smallpox-virus>.

Chapter 8 Information warfare

Ahmed N. Climate Science Denial Network Behind Great Barrington Declaration. *Byline Times*. October 9, 2020, <https://bylinetimes.

com/2020/10/09/climate-science-denial-network-behind-great-barrington-declaration/>.

Andrejko KL, Pry JM, Myers JF, et al. Effectiveness of Face Mask or Respirator Use in Indoor Public Settings for Prevention of SARS-CoV-2 Infection – California, February–December 2021. *Morbidity and Mortality Weekly Report.* 2022, 71: 212–16.

Brusselaers N, Steadson D, Bjorklund K, Breland S, Stilhoff Sörensen J, Ewing A, et al. Evaluation of Science Advice during the COVID-19 Pandemic in Sweden. *Nature Humanities and Social Sciences Communications.* 2022, 9(1): 91.

Cohen J. Covid-19 Fallout: Ruinous Effects of Politicization of Public Health Agencies, Such as the CDC. *Forbes.* April 2022, <www.forbes.com/sites/joshuacohen/2022/04/01/covid-19-fallout-ruinous-effects-of-politicization-of-public-health-agencies-such-as-the-cdc/>.

Dr Yak. Challenging Science's Status-Quo: The Tale of Barry Marshall. *Medium.* July 2019, <https://medium.com/doctoryak/challenging-sciences-status-quo-the-tale-of-barry-marshall-c80a873412a6>.

Heywood AE, and MacIntyre CR. Elimination of COVID-19: What Would It Look Like and Is It Possible? *Lancet Infectious Diseases.* September 2020, 20(9): 1005–1007.

Hotez PJ. Anti-Science Kills: From Soviet Embrace of Pseudoscience to Accelerated Attacks on US Biomedicine. *PLoS Biol.* January 28, 2021, 19(1): e3001068.

Jewish Virtual Library. Joseph Goebbels: On the 'Big Lie', <www.jewishvirtuallibrary.org/joseph-goebbels-on-the-quot-big-lie-quot>.

Klompas M, Diekema DJ, Fishman NO, and Yokoe DS. Ebola Fever: Reconciling Ebola Planning with Ebola Risk in U.S. Hospitals. *Annals of Internal Medicine.* November 18, 2014, 161(10): 751–52.

Langer WC. A Psychological Analysis of Adolph Hitler: His Life and Legend. MO Branch, Office of Strategic Services, Washington, DC, <www.cia.gov/readingroom/docs/CIA-RDP78-02646R000600240001-5.pdf>.

Lawrence L. Cigarettes Were Once 'Physician' Tested, Approved. *Healio*, March 2009, <www.healio.com/news/hematology-oncology/20120325/cigarettes-were-once-physician-tested-approved>.

MacIntyre CR, Chughtai AA, Seale H, Richards GA, and Davidson PM. Respiratory Protection for Healthcare Workers Treating Ebola Virus Disease (EVD): Are Facemasks Sufficient to Meet Occupational Health and Safety Obligations? *International Journal of Nursing Studies.* 2014, 50(11): 1421–26.

Martin-Moreno JM, Llinás G, and Hernández JM. Is Respiratory Protection Appropriate in the Ebola Response? *Lancet.* September 6, 2014, 384(9946): 856.

Piplani S, Singh PK, Winkler DA, and Petrovsky N. In Silico Comparison of SARS-CoV-2 Spike Protein-ACE2 Binding Affinities Across Species and Implications for Virus Origin. *Scientific Reports.* June 24, 2021, 11(1): 13063.

The Ventilation Revolution, <www.youtube.com/watch?v=nWnOR3O-ZF0>.

References

Chapter 9 Trust me, I'm a doctor

Bolsin SN. Professional Misconduct: The Bristol Case. *Medical Journal of Australia.* October 5, 1998, 169(7): 369–72.

Davies HT, and Shields AV. Public Trust and Accountability for Clinical Performance: Lessons from the National Press Reportage of the Bristol Hearing. *Journal of Evaluation in Clinical Practice.* August 1999, 5(3): 335–42.

Goad, J. Killer Doctors: 8 Physicians Who Murdered Their Patients. *Thought Catalog.* June 2021, <https://thoughtcatalog.com/jim-goad/2020/01/killer-doctors-8-physicians-who-murdered-their-patients/>.

Hanauske-Abel HM. Not a Slippery Slope or Sudden Subversion: German Medicine and National Socialism in 1933. *BMJ.* 1996, 313(7070): 1453–63.

Mbali M. 'A Matter of Conscience': The Moral Authority of the World Medical Association and the Readmission of the South Africans, 1976–1994. *Medical History.* April 2014, 58(2): 257–77.

Open Letter. Free Speech for Doctors. Letter to Health Ministers: Re Ahpra and Dr David Berger. 14 July 2022, <https://speakupdoc.com/>.

Palm Island Verdict: 12 Years in the Making. *SBS News.* December 2016D, <www.sbs.com.au/news/article/palm-island-verdict-12-years-in-the-making/7e1tjirfc>.

Sanggaran JP, Haire B, and Zion D. The Health Care Consequences of Australian Immigration Policies. *PLoS Med.* February 16, 2016, 13(2): e1001960.

Silove D. Doctors and the State: Lessons from the Biko Case. *Social Science & Medicine.* 1990, 30(4): 417–29.

Swannell, C. AMA Victoria to Call for Royal Commission into AHPRA. *MJA Insight.* July 2022, 27, <https://insightplus.mja.com.au/2022/27/ama-victoria-to-call-for-royal-commission-into-ahpra/>.

Swannell, C. Fallout Continues from AHPRA 'Over-Reach'. *MJA Insight.* August 2022, 29, <https://insightplus.mja.com.au/2022/29/fallout-continues-from-ahpra-over-reach/>.

Wiesel E. Without Conscience. *New England Journal of Medicine.* 2005, 352(15): 1511–13.

Chapter 10 Epidemic detectives

Awadh A, Chughtai AA, Dyda A, Sheikh M, Heslop DJ, and MacIntyre CR. Does Zika Virus Cause Microcephaly – Applying the Bradford Hill Viewpoints. *PLoS Currents.* February 22, 2017, 9.

Centers for Disease Control and Prevention. CDC's Origins and Malaria, <https://www.cdc.gov/malaria/about/history/history_cdc.html>.

Chen X, Chughtai AA, and MacIntyre CR. A Systematic Review of Risk Analysis Tools for Differentiating Unnatural from Natural Epidemics. *Military Medicine.* November 2017, 182(11): e1827-e1835.

Chen X, Chughtai AA, and MacIntyre CR. Recalibration of the Grunow-Finke Assessment Tool to Improve Performance in Detecting Unnatural Epidemics. *Risk Analysis.* July 2019, 39(7): 1465–75.

Chow A. I Spent 80 Minutes Inside Vitalik Buterin's Brain. Here's What I Learned. *Time*. March 2022, <https://time.com/6157862/vitalik-buterin-interview-transcript/>.

Di Cicco ME, Ragazzo V, and Jacinto T. Mortality in Relation to Smoking: The British Doctors Study. *Breathe*. 2016, 12(3): 275–76.

Dyda A, Stelzer-Braid S, Adam D, Chughtai AA, and MacIntyre CR. The Association Between Acute Flaccid Myelitis (AFM) and Enterovirus D68 (EV-D68) – What is the Evidence for Causation? *Eurosurveillance*. 2018, 23(3): 17-00310.

Epidemic Intelligence Service. A History of Success: Investigating and Responding to Public Health Threats Since 1951, <https://www.cdc.gov/eis/about/history.html>.

John Snow's Pump (1854). British Society for Immunology, <www.immunology.org/john-snows-pump-1854>.

Joshi A, Sparks R, Karimi S, Yan SJ, Chughtai AA, Paris C, and MacIntyre CR. Automated Monitoring of Tweets for Early Detection of the 2014 Ebola Epidemic. *PLoS One*. March 17, 2020, 15(3): e0230322.

Joshi A, Sparks R, McHugh J, Karimi S, Paris C, and MacIntyre CR. Harnessing Tweets for Early Detection of an Acute Disease Event. *Epidemiology*. January 2020, 31(1): 90–97.

Kpozehouen EB, Chen X, Zhu M, and MacIntyre CR. Using Open-Source Intelligence to Detect Early Signals of COVID-19 in China: Descriptive Study. *JMIR Public Health Surveillance*. September 18, 2020, 6(3): e18939.

MacIntyre CR. Biopreparedness in the Age of Genetically Engineered Pathogens and Open Access Science: An Urgent Need for a Paradigm Shift. *Military Medicine*. September 2015, 180(9): 943–49.

MacIntyre CR, Doolan C, and De Silva C. The Explosion at Vector: Hoping for the Best While Preparing for the Worst. *Global Biosecurity*. 2019, 1(1).

Pendergrast M. *Inside the Outbreaks: The Elite Medical Detectives of the Epidemic Intelligence Service*. Houghton Mifflin Harcourt, New York, 2010.

Quigley AL, Stone H, Nguyen PY, Chughtai AA, and MacIntyre CR. Estimating the Burden of COVID-19 on the Australian Healthcare Workers and Health System during the First Six Months of the Pandemic. *International Journal of Nursing Studies*. February 2021, 114: 103811.

Schultz MG, and Schaffner W. Alexander Duncan Langmuir. *Emerging Infectious Diseases*. 2015, 21(9): 1635–37.

Thamtono Y, Moa A, and MacIntyre CR. Using Open-Source Intelligence to Identify Early Signals of COVID-19 in Indonesia. *Western Pacific Surveillance and Response Journal*. February 17, 2021, 12(1): 40–45.

References

Chapter 11 The fuss about facemasks

Andrejko KL, Pry JM, Myers JF, et al. Effectiveness of Face Mask or Respirator Use in Indoor Public Settings for Prevention of SARS-CoV-2 Infection – California, February–December 2021. *Morbidity and Mortality Weekly Report*. 2022, 71: 212–16.

Bahl P, Bhattacharjee S, De Silva C, Chughtai AA, Doolan C, and MacIntyre CR. Face Coverings and Mask to Minimise Droplet Dispersion and Aerosolisation: A Video Case Study. *Thorax*. 2020, 75: 1024–25.

Jaax J, and Jaax NK. An Ebola Filovirus Is Discovered in the USA: Reston, Virginia, USA, 1989. *Veterinary Heritage*. May 2016, 39(1): 16–19.

Klompas M, Diekema DJ, Fishman NO, and Yokoe DS. Ebola Fever: Reconciling Ebola Planning with Ebola Risk in U.S. Hospitals. *Annals of Internal Medicine*. November 18, 2014, 161(10): 751–52.

MacIntyre CR, Ananda-Rajah M, Nicholls M, and Quigley AL. Current COVID-19 Guidelines for Respiratory Protection of Health Care Workers Are Inadequate. *Medical Journal of Australia*. September 2020, 213(6): 251–52e1.

MacIntyre CR, and Ananda-Rajah MR. Scientific Evidence Supports Aerosol Transmission of SARS-CoV-2. *Antimicrobial Resistance & Infection Control*. December 18, 2020, 9(1): 202.

MacIntyre CR, and Wang Q. Physical Distancing, Face Masks, and Eye Protection for Prevention of COVID-19. *The Lancet*. June 27, 2020, 395(10242): 1973–87.

MacIntyre CR, Cauchemez S, Dwyer DE, Seale H, Cheung P, Browne G, et al. Face Mask Use and Control of Respiratory Virus Transmission in Households. *Emerging Infectious Diseases*. 2009, 15(2): 233–41.

MacIntyre CR, Chughtai A, Bhattacharjee S, Kunasekaran M, and Engells T. Risk Mitigation of Inadvertent Exposure to Biothreats to Front Line Law Enforcement. *Global Biosecurity*. 2020, 2(1).

MacIntyre CR, Chughtai AA, Rahman B, Peng Y, Zhang Y, Seale H, et al. The Efficacy of Medical Masks and Respirators Against Respiratory Infection in Healthcare Workers. *Influenza and Other Respiratory Viruses*. 2017, 11(6): 511–17.

MacIntyre CR, Chughtai AA, Seale H, Richards GA, and Davidson PM. Respiratory Protection for Healthcare Workers Treating Ebola Virus Disease (EVD): Are Facemasks Sufficient to Meet Occupational Health and Safety Obligations? *International Journal of Nursing Studies*. November 2014, 51(11): 1421–26.

MacIntyre CR, Costantino V, and Kunasekaran MP. Health System Capacity in Sydney, Australia in the Event of a Biological Attack with Smallpox. *PLoS One*. June 14, 2019, 14(6): e0217704.

MacIntyre CR, Dung TC, Chughtai AA, Seale H, and Rahman B. Contamination and Washing of Cloth Masks and Risk of Infection among

Hospital Health Workers in Vietnam: A Post Hoc Analysis of a Randomised Controlled Trial. *BMJ Open*. 2020, 10(9): e042045.

MacIntyre CR, Seale H, Dung TC, Hien NT, Nga PT, Chughtai AA, et al. A Cluster Randomised Trial of Cloth Masks Compared with Medical Masks in Healthcare Workers. *BMJ Open*. 2015, 5(4): e006577-e.

MacIntyre CR, Wang Q, Cauchemez S, Seale H, Dwyer DE, Yang P, et al. A Cluster Randomized Clinical Trial Comparing Fit-Tested and Non-Fit-Tested N95 Respirators to Medical Masks to Prevent Respiratory Virus Infection in Health Care Workers. *Influenza and Other Respiratory Viruses*. 2011, 5(3): 170–79.

MacIntyre CR, Wang Q, Rahman B, Seale H, Ridda I, Gao Z, et al. Efficacy of Face Masks and Respirators in Preventing Upper Respiratory Tract Bacterial Colonization and Co-Infection in Hospital Healthcare Workers. *Preventive Medicine*. 2014, 62: 1–7.

MacIntyre CR, Wang Q, Seale H, Yang P, Shi W, Gao Z, et al. A Randomized Clinical Trial of Three Options for N95 Respirators and Medical Masks in Health Workers. *American Journal of Respiratory and Critical Care Medicine*. 2013, 187(9): 960–66.

MacIntyre CR, Zhang Y, Chughtai AA, Seale H, Zhang D, Chu Y, et al. Cluster Randomised Controlled Trial to Examine Medical Mask Use as Source Control for People with Respiratory Illness. *BMJ Open*. 2016, 6(12): e012330-e.

Martin-Moreno JM, Llinás G, and Hernández JM. Is Respiratory Protection Appropriate in the Ebola Response? *Lancet*. September 6, 2014, 384(9946): 856.

Chapter 12 Brain eating viruses

Brusselaers N, Steadson D, Bjorklund K, et al. Evaluation of Science Advice during the COVID-19 Pandemic in Sweden. *Nature Humanities and Social Sciences Communications*. 2022, 9(1): 91.

Chertow D, Stein S, Ramelli S, et al. SARS-CoV-2 Infection and Persistence Throughout the Human Body and Brain, 20 December 2021, Preprint (Version 1) available at <https://assets.researchsquare.com/files/rs-1139035/v1_covered.pdf?c=1640020576>.

Douaud G, Lee S, Alfaro-Almagro F. et al. SARS-CoV-2 is Associated with Changes in Brain Structure in UK Biobank. *Nature*. 2022, 604: 697–707.

Fearnley KA. Through My Eyes: Long Neuro-COVID. *Medical News Today*. November 2021, <www.medicalnewstoday.com/articles/through-my-eyes-long-neuro-covid>.

Fernández-Castañeda A, Lu P, Geraghty AC, et al. Mild Respiratory SARS-CoV-2 Infection Can Cause Multi-Lineage Cellular Dysregulation and Myelin Loss in the Brain. *bioRxiv* [Preprint]. January 10, 2022: 2022.01.07.475453.

Hampshire A, Trender W, Chamberlain SR, Jolly AE, et al. Cognitive Deficits in People Who Have Recovered from COVID-19. *Lancet eClinicalMedicine*. 2021, 39.

Horby P. Variant Creutzfeldt-Jakob Disease: An Unfolding Epidemic of Misfolded Proteins. *Journal of Paediatric Child Health*. December 2002, 38(6): 539–42.

Maury A, Lyoubi A, Peiffer-Smadja N, de Broucker T, and Meppiel E. Neurological Manifestations Associated With SARS-Cov-2 And Other Coronaviruses: A Narrative Review for Clinicians. *Revue Neurologique*. January–February 2021, 177(1–2): 51–64.

Pilotto A, Masciocchi S, Volonghi I, et al. SARS-CoV-2 Related Encephalopaties (ENCOVID) Study Group. Clinical Presentation and Outcomes of Severe Acute Respiratory Syndrome Coronavirus 2-Related Encephalitis: The ENCOVID Multicenter Study. *Journal of Infectious Diseases*. January 4, 2021, 223(1): 28–37.

Qureshi AI, Baskett WI, Huang W, et al. New-Onset Dementia Among Survivors of Pneumonia Associated with Severe Acute Respiratory Syndrome Coronavirus 2 Infection, *Open Forum Infectious Diseases*, April 2022, 19(4): ofac115.

Shen WB, Logue J, Yang P, Baracco L, et al. SARS-CoV-2 Invades Cognitive Centers of the Brain and Induces Alzheimer's-Like Neuropathology. *bioRxiv* [Preprint]. February 3, 2022: 2022.01.31.478476.

Smith C, Odd D, Harwood R, et al. Deaths in Children and Young People in England after SARS-CoV-2 Infection during the First Pandemic Year. *Nature Medicine*. 2022, 28: 185–92.

Xie Y, Xu E, Bowe B, et al. Long-Term Cardiovascular Outcomes Of COVID-19. *Nature Medicine*. 2022, 28: 583–90.

Chapter 13 The biological sniper weapon

Almuzaini T, Sammons H, and Choonara I. Substandard and Falsified Medicines in the UK: A Retrospective Review of Drug Alerts (2001–2011). *BMJ Open*. July 24, 2013, 3(7): e002924.

Andrade GE, and Hussain A. Polio in Pakistan: Political, Sociological, and Epidemiological Factors. *Cureus*. October 27, 2018, 10(10): e3502.

Bellingcat Investigation Team. Full report: Skripal Poisoning Suspect Dr. Alexander Mishkin, Hero of Russia. *Bellingcat*. October 2018, <www.bellingcat.com/news/uk-and-europe/2018/10/09/full-report-skripal-poisoning-suspect-dr-alexander-mishkin-hero-russia/>.

Blachere FM, Lindsley WG, Pearce TA, Anderson SE, Fisher M, Khakoo R, et al. Measurement of Airborne Influenza Virus in a Hospital Emergency Department. *Clinical Infectious Diseases*. 2009, 48(4): 438–40.

Brooks L. Valium Recall: Consumers Urged to Check Valpam 5 Packs After Suspected Tampering. *ABC News*. June 2017, <www.abc.net.au/news/2017-06-10/valium-recall-affects-valpam-5/8606646>.

Cousins S. Contaminated Vaccine Deaths a Serious Setback for Syria. *Lancet*. September 27, 2014, 384(9949): 1172.

Dershewitz RA, and Levin GS. The Effect of the Tylenol Scare on Parent's Use of Over-the-Counter Drugs. *Clinical Pediatrics*. August 1984, 23(8): 445–48.

Douglas RM, Hemilä H, Chalker E, and Treacy B. Vitamin C for Preventing and Treating the Common Cold. *Cochrane Database of Systematic Reviews*. July 18, 2007, 3: CD000980.

Figueroa JM, Lombardo ME, Dogliotti A, Flynn LP, Giugliano R, Simonelli G, Valentini R, Ramos A, Romano P, Marcote M, Michelini A, Salvado A, Sykora E, Kniz C, Kobelinsky M, Salzberg DM, Jerusalinsky D, and Uchitel O. Efficacy of a Nasal Spray Containing Iota-Carrageenan in the Postexposure Prophylaxis of COVID-19 in Hospital Personnel Dedicated to Patients Care with COVID-19 Disease. *International Journal of General Medicine*. October 1, 2021, 14: 6277–86.

Getlen L. How a Murderous Doctor Was Allowed to Keep Killing Patients. *New York Post*. September 2018, <https://nypost.com/2018/09/08/how-a-murderous-doctor-was-allowed-to-keep-killing-patients/>.

Graff GM. China's 5 Steps for Recruiting Spies. Cases of Americans Allegedly Recruited to Spy on China's Behalf Follow a Basic Pattern. *Wired*. October 2018, <www.wired.com/story/china-spy-recruitment-us/>.

Harrison J, Fell T, Leggett R, Lloyd D, Puncher M, and Youngman M. The Polonium-210 Poisoning of Mr Alexander Litvinenko. *Journal of Radiological Protection*. March 20, 2017, 37(1): 266–78.

Health Hazard Alerts. Suspected Tampering Associated with Various Ham and Sliced Meat Products. Canadian Food Inspection Agency, Ottawa, Ontario. *Canada Communicable Disease Report*. December 15, 2006, 32(24): 297.

Hemilä H. Common Cold Treatment Using Zinc. *Journal of the American Medical Association*. August 18, 2015, 314(7): 730.

Israeli Research Shows Medical Scans Vulnerable to Hacking. *Medical Device Network*. April 2019, <www.medicaldevice-network.com/news/medical-scans-cybersecurity-study/>.

Jalali MS, Razak S, Gordon W, Peraklis E, and Madnick S. Health Care and Cybersecurity: Bibliometric Analysis of the Literature. *Journal of Medical Internet Research*. February 15, 2019, 21(2): e12644.

Kitajima M, Huang Y, Watanabe T, Katayama H, and Haas CN. Dose-Response Time Modelling for Highly Pathogenic Avian Influenza A (H5N1) Virus Infection. *Letters in Applied Microbiology*. October 2011, 53(4): 438–44.

Lednicky JA, Lauzardo M, Fan ZH, Jutla A, Tilly TB, Gangwar M, et al. Viable SARS-CoV-2 in the Air of a Hospital Room with COVID-19 Patients. *International Journal of Infectious Diseases*. 2020, 100: 476–82.

Lissiman E, Bhasale AL, and Cohen M. Garlic for the Common Cold. *Cochrane Database of Systematic Reviews*. November 11, 2014, 11: CD006206.

MacIntyre CR, and Bui CM. Pandemics, Public Health Emergencies and Antimicrobial Resistance – Putting the Threat in an Epidemiologic and Risk Analysis Context. *Archives of Public Health*. September 14, 2017, 75: 54.

References

MacIntyre CR, and Chughtai AA. Recurrence and Reinfection: A New Paradigm for the Management of Ebola Virus Disease. *International Journal of Infectious Diseases*. February 2016, 43: 58–61.

MacIntyre CR, Engells TE, Scotch M, Heslop DJ, Gumel AB, Poste G, et al. Converging and Emerging Threats To Health Security. *Environment Systems and Decisions*. 2018, 38(2): 198–207.

Manning L, Baines RN, and Chadd SA. Deliberate Contamination of the Food Supply Chain. *British Food Journal*. 2005, 107, 225–45.

Mullaney A, and Hassan SA. He Led the CIA to bin Laden – and Unwittingly Fueled a Vaccine Backlash. National Geographic. February 2015, <www.nationalgeographic.com/science/article/150227-polio-pakistan-vaccination-taliban-osama-bin-laden>.

Nemhauser JB. The Polonium-210 Public Health Assessment: The Need for Medical Toxicology Expertise in Radiation Terrorism Events. *Journal of Medical Toxicology*. September 2010, 6(3): 355–59.

Oduwole O, Udoh EE, Oyo-Ita A, and Meremikwu MM. Honey for Acute Cough in Children. *Cochrane Database of Systematic Reviews*. April 10, 2018, 4(4): CD007094.

Pesticide-Laced Salt at Restaurant Sickened 107, Stumped Investigators. *CIDRAP News*. August 2002, <www.cidrap.umn.edu/news-perspective/2002/08/pesticide-laced-salt-restaurant-sickened-107-stumped-investigators>.

Rosenbaum E. 5 Biggest Risks of Sharing Your DNA with Consumer Genetic-Testing Companies. *CNBC Disruptor 50*. June 2018, <www.cnbc.com/2018/06/16/5-biggest-risks-of-sharing-dna-with-consumer-genetic-testing-companies.html>.

Rule A. *Bitter Harvest*. Simon & Schuster, New York, 1999.

Safi M, and Baloch SM. Pakistan Doctor Held After 437 Children Diagnosed with HIV. *The Guardian*. May 2019, <www.theguardian.com/world/2019/may/17/pakistan-doctor-held-after-400-children-diagnosed-with-hiv>.

Samoa Court to Probe Infant Deaths. *RNZ*. August 2019, <www.rnz.co.nz/international/pacific-news/395890/samoa-court-to-probe-infant-deaths>.

Shah S. CIA Organised Fake Vaccination Drive to Get Osama Bin Laden's Family DNA. *The Guardian*. July 2011, <www.theguardian.com/world/2011/jul/11/cia-fake-vaccinations-osama-bin-ladens-dna>.

Sommersguter-Reichmann M, Wild C, Stepan A, Reichmann G, and Fried A. Individual and Institutional Corruption in European and US Healthcare: Overview and Link of Various Corruption Typologies. *Applied Health Economics and Health Policy*. June 2018, 16(3): 289–302.

Taylor GJ. Physician as Serial Killer. *New England Journal of Medicine*. July 28, 2005, 353(4): 430.

Vranka MA, and Bahník Š. Predictors of Bribe-Taking: The Role of Bribe Size and Personality. *Frontiers in Psychology*. September 10, 2018, 9: 1511.

Wagner L, Cramer H, Klose P, Lauche R, Gass F, Dobos G, and Langhorst J. Herbal Medicine for Cough: A Systematic Review and Meta-Analysis. *Forsch Komplementmed*. 2015, 22(6): 359–68.

Woman Charged Over Sizzler Rat Poison Scare. *The Age*. March 2006, <www.theage.com.au/national/woman-charged-over-sizzler-rat-poison-scare-20060302-ge1uw1.html>.

Chapter 14 A biological winter

AAP. Man Who Implanted Opal Travel Card Chip Has Conviction Overturned. *The Guardian*. June 2018, <www.theguardian.com/australia-news/2018/jun/18/man-who-implanted-opal-travel-card-chip-has-conviction-overturned>.

Brusselaers N, Steadson D, Bjorklund K, et al. Evaluation of Science Advice during the COVID-19 Pandemic in Sweden. *Nature Humanities and Social Sciences Communications*. 2022, 9(1): 91.

DIYbio. Find a Biohacking Makerspace Near You. *Makezine*. April 2017, <https://makezine.com/2017/04/05/biohacking-spaces/>.

Harrison NL, and Sachs JD. A Call for an Independent Inquiry into the Origin of the SARS-CoV-2 Virus. *Proceedings of the National Academy of Sciences of the United States of America*. May 24, 2022, 119(21): e2202769119.

Heslop DJ, and MacIntyre CR, 2019. Germ Line Genome Editing and the Emerging Struggle For Supremacy in the Chemical, Biological and Radiological (CBR) Balance of Power. *Global Biosecurity*. 2019, 1(1), <https://jglobalbiosecurity.com/article/10.31646/gbio.18/>.

Kaminska I. The Garage Biohackers Who Manipulate DNA. *Australian Financial Review*, <www.afr.com/companies/healthcare-and-fitness/the-garage-biohackers-who-manipulate-dna-20210921-p58tmn>.

MACE E PAI COVID 19 ANALYSIS Redacted. Contributed by *NBC News Investigations* (NBC). 2020, <www.documentcloud.org/documents/6884792-MACE-E-PAI-COVID-19-ANALYSIS-Redacted.html>.

McCaughey B. Lessons from SARS: A Tale of Two Cities. *Modern Healthcare*. February 2007, <www.modernhealthcare.com/article/20070216/NEWS/70216020/lessons-from-sars-a-tale-of-two-cities>.

McCormack W. Outside the Limits of the Human Imagination: What the New Documentary 'Wild, Wild Country' Doesn't Capture about the Magnetism and Evil of the Rajneesh Cult. *New Republic*. March 2018, <https://newrepublic.com/article/147657/outside-limits-human-imagination>.

Rabia, YP. From Bioweapons To Super Soldiers: How the UK is Joining the Genomic Technology Arms Race. *The Conversation*. April 2021, <https://theconversation.com/from-bioweapons-to-super-soldiers-how-the-uk-is-joining-the-genomic-technology-arms-race-159889>.

Robinson N. Why the Chair of the Lancet's COVID-19 Commission Thinks the US Government Is Preventing a Real Investigation into the Pandemic.

References

Current Affairs. August 2022, <www.currentaffairs.org/2022/08/why-the-chair-of-the-lancets-covid-19-commission-thinks-the-us-government-is-preventing-a-real-investigation-into-the-pandemic>.

US Food and Drug Administration. Human Gene Therapy Products Incorporating Human Genome Editing. March 2022, <www.fda.gov/regulatory-information/search-fda-guidance-documents/human-gene-therapy-products-incorporating-human-genome-editing>.

World Health Organization. Human Genome Editing: A Framework for Governance and Recommendations, July 14, 2021, <www.who.int/publications/i/item/9789240030381>.

Milton Keynes UK
Ingram Content Group UK Ltd.
UKHW042133031023
429886UK00004BA/236